Advanced Laser and Competing Technologies Easily Explained

Advanced Laser and Competing Technologies Easily Explained

Dieter Schuöcker
Technical University of Vienna, Austria

Georg Schuöcker

World Scientific

NEW JERSEY · LONDON · SINGAPORE · BEIJING · SHANGHAI · HONG KONG · TAIPEI · CHENNAI · TOKYO

Published by

World Scientific Publishing Co. Pte. Ltd.
5 Toh Tuck Link, Singapore 596224
USA office: 27 Warren Street, Suite 401-402, Hackensack, NJ 07601
UK office: 57 Shelton Street, Covent Garden, London WC2H 9HE

Library of Congress Control Number: 2021949869

British Library Cataloguing-in-Publication Data
A catalogue record for this book is available from the British Library.

Cover image: Laser cladding (with kind permission of Trumpf)
Inset: 3D printing with a laser (Source: Authors, at TU Vienna)

ADVANCED LASER AND COMPETING TECHNOLOGIES EASILY EXPLAINED

ISBN 978-981-124-635-7 (hardcover)
ISBN 978-981-124-636-4 (ebook for institutions)
ISBN 978-981-124-637-1 (ebook for individuals)

For any available supplementary material, please visit
https://www.worldscientific.com/worldscibooks/10.1142/12529#t=suppl

Desk Editor: Joseph Ang

Typeset by Stallion Press
Email: enquiries@stallionpress.com

The authors dedicate this book to their descendants:
Nils and Lars Mosser; Madita and Marlies Ilvy Schuöcker.

Introductory Remarks

Many years ago one of the authors worked in the industry and developed a laser cutting and punching machine. Once he proposed to the CEO to visit his development, but the director answered that he does not like to see the laser system because he does not understand it and so the authors had the idea to write the actual book since the use of lasers in production engineering is quite advantageous but obstructed many times by a poor understanding of physics underlying laser technology, as optics, plasma physics, heat conduction and others. Nevertheless laser technology is not so difficult to understand as it seems if those physical phenomena are explained in simple words generally understandable without mathematical derivations and other theoretical stuff, being the aim of this book. In order to ease understanding even more, for many phenomena treated in the book plausibility models have been developed, that use strong simplifications, but yield correct results compared to experimental findings.

The book is divided in two main parts, one on lasers and one on mechanical, electrical and thermal technologies, allowing a comparison and evaluation. Each part begins with tutorials, the first on basics of laser technology and the second on fundamentals of mechanical, electrical and thermal technology. The first tutorials are followed by five chapters, the first one dealing with all the equipment needed for laser materials processing, as the most important laser sources, beam forming, transport (fibers) and direction (scanners) and also sensors especially for the temperature of the workpiece. The next three chapters are devoted to laser manufacturing processes with material loss as laser cutting, material gain as welding and

3D printing with lasers and unchanged mass as hardening or bending brittle materials with laser assistance. In these chapters the main process mechanisms and performance as well as newly achieved processes are treated. In a fifth chapter, safety for personal and environment, a most important topic, is treated in detail.

In the part on competitive technologies, after the tutorials as mentioned above on mechanical, electrical and thermal technologies three chapters are devoted to the latter conventional manufacturing processes. Unlike to the laser section, the necessary equipment is not treated in a separate chapter but is integrated in the process chapters, since the various conventional production procedures use individual tools and not more or less equal machinery as for the most important laser processes. Similar to the part on laser processing, manufacturing with mass loss as for instance machining, then processes with mass gain as for instance welding and finally hardening without change of workpiece mass are treated. Also machine safety is subject of a further chapter.

In a last chapter the laser processes and the conventional ones are compared, finally leading to a general judgment on the importance of the various laser solutions and their future prospects.

Hopefully the actual book will serve to interested people that are not familiar with lasers to get a deeper understanding of laser techniques and their advantages and shortcomings in view of conventional technologies. Moreover it should present the most recent progress on laser technology, thus serving even people active in lasers and production technology to stay up to date.

Acknowledgments

The authors appreciate gratefully the kind permission of the monthly journal METALL, published by Österreichischer Wirtschaftsverlag Wien, to use material of LaserNews for the actual book.

The senior author also appreciates the most valuable developments of his former coworkers Markus Bohrer (coaxial laser), Andreas Penz (laser plotter/Trotec), Alexander Kratky (3D printing of metals), Ferdinand

Bammer (laser assisted bending), Richard Majer (FEM simulations) and Joachim Aichinger (laser hardening with temperature control).

Dieter Schuöcker[1]
Georg Schuöcker[2]

---

[1] Dieter Schuöcker (senior author) is Professor Emeritus at TU Vienna, Production and Photonics Technologies.
[2] Georg Schuöcker is Head of Production at a precision machining company.

Contents

Chapter 1

Light, Lasers and Laser Heating (Tutorial)

1.1 Waves in General and Light Waves

If a stone hits the initially calm surface of water a dip appears and water is displaced and forms a little circular wall around the tip being called a wave. The latter moves away in all directions with a certain speed and expands to form larger circles. If now after a certain time another stone hits the water surface a second wave appears and also starts to move in all directions. In that moment, the first wave has reached a certain distance. Thus a momentary view shows that two wave crests can be seen with a distance given by the time between two hitting events multiplied by the speed of propagation. The time between two distortions at the origin of the wave is the *period length* and the distance that the first wave travelled in that time is the *wavelength*. If the hits appear at the origin permanently with a frequency given by the reciprocal period length, waves leave one after the other, forming concentric circles separated by the wavelength.

Quite similar a distortion in a point of the initially electrically neutral space, say by the immediate deliberation of a negatively charged electron from an atom, leaving a positive ion, results in the sudden appearance of an electrical force. The latter causes a wave of the field strength propagating in all directions. Due to the natural law that electrical fields with time dependent strength cause always the appearance of magnetic forces [1], so called *electromagnetic waves* consist of an electrical field strength associated by a perpendicular magnetic field. If the human eye can register such waves, usually at wavelengths between 0.4 μm (violet) and 0.75 μm (red), the wave is called *visible light*.

Natural light sources are the sun but also lamps and similar devices. These sources emit light as wave packages that have a finite length, are independent of each other and travel in all directions.

In contrary artificial light generated by lasers has a continuous wave train emitted in a distinct direction and is called *coherent light*.

1.2 Diffraction of Light at a Narrow Slot and Gaussian Beam

If a narrow slot is illuminated, the waves propagating along the axis (see Fig. 1.1) are transmitted as the main lobe. Nevertheless the edges of the slot also hit by the light wave emitt themselves according to Huygen's law [1] also light in all directions, where in a certain direction-inclined towards the axis-the light emitted by the lower edge and that of the upper edge show a travelling way difference of a wavelength, thus yielding coincident maxima and mutual enhancement causing a first side lobe (Fig. 1.1). Subsequent side lobes are obtained for way differences of a multiple of the wavelength, where the intensity becomes weaker the more a lobe is apart from the main lobe, since most of the energy flows through the inner lobes. If now the slot becomes wider, the first side lobe and all the others come nearer to the main one (see Fig. 1.1) and finally for very large macroscopic slots they touch each other and a continuous decreasing brightness of the light transmitted through the slot is obtained. With rising distance from the slot, the lobes separate more and more from each other and therefore the light distribution becomes wider.

The latter considerations are very useful to understand the properties of a beam of light coming from a source with finite extension, where the latter source can be regarded as a large, circular opening comparable to a wide slot as treated above. Due to the above phenomena the beam must have a cross section that resembles a bell or *Gaussian* curve with a maximum in the axis and a continuous decrease in radial direction till infinity. Accordingly a beam radius and width

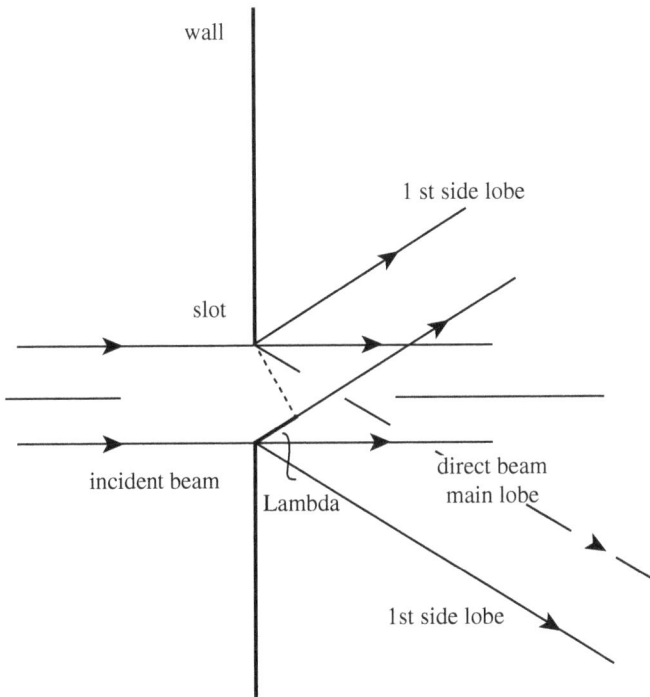

Fig. 1.1 Diffraction at a narrow slot.

Source: Authors.

cannot be determined. Thus a nominal radius must be defined and is allocated at a radius where the *intensity of the beam* (energy crossing unit area) is reduced to 10% of the maximum. The angle of the first side lobe is given by the wavelength divided by the slot width (see Fig. 1.2) and can be regarded as the opening angle of the beam transmitted by the slot and is called *divergence*.

Finally a Gaussian beam is defined by its nominal radius or the *width* and the opening angle, the angle of the first side lobe, is called the *divergence*. The product of both yields according to the above considerations for a slot the wavelength or for the most important geometry of a circular hole a fraction of the wavelength (λ/π) [2, 3]. Since the actual product depends only on the wavelength (more correct also on the beam quality in terms of focusability) it remains always unchanged whatever happens to

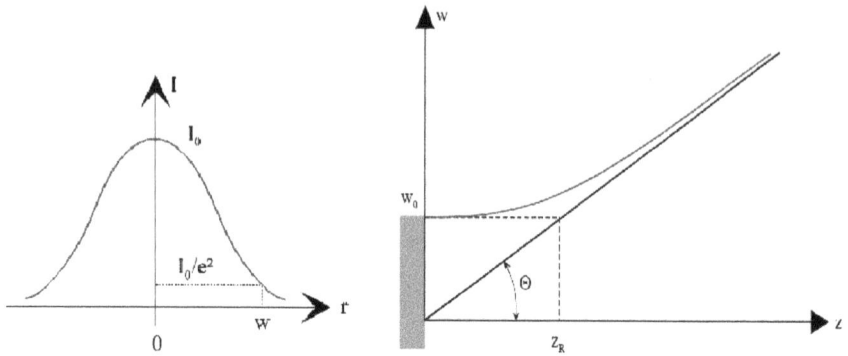

Fig. 1.2   (Left) Radial intensity distribution, (Right) beam edge.

Source: D. Schuöcker, Spanlose Fertigung, Fig. 1.2.3 and 1.2.5 Oldenbourg, München 2004.

the beam. Examples for the latter product are 0.3 mm mrad for a Gaussian beam with wavelength around 1 μm, about 10 mm mrad for a realistic solid state laser and around 100 mm mrad for a realistic CO_2 laser.

Besides the Gaussian beam there are also beam modes with a main lobe and side lobes. They are unwanted because they allow only imperfect focusing to a central spot and satellite rings and can be avoided in lasers by a geometry with small mirrors separated by a large distance, because due to the beam divergence a large part of the beam cross section including the side lobes is cut out by the mirrors, thus allowing only the main lobe to be amplified [2].

1.3   Reflection and Refraction of Light Beams

If a light wave hits the atoms in the bulk of a glass the electrons are pushed slightly in one direction by the electrical field what leads to the appearance of negative charges on the surface of the atom on one side and of positive charges on the side where electrons have been pushed away. Nevertheless the same happens to the neighboring atoms, thus compensating the apparent surface charges. Only at the surface of the glass no atoms exist outside and so complete compensation of the surface charges is not possible and the atoms show charges with a polarity

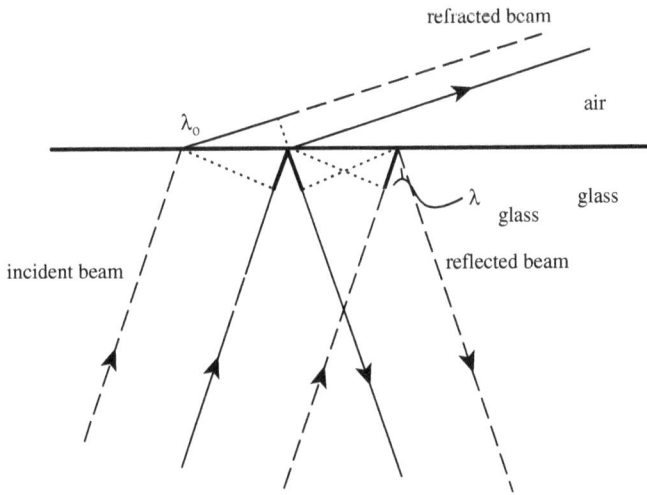

Fig. 1.3 Reflection and refraction.

Source: Authors.

that alternates with light frequency. Thus, an oscillating electrical field strength appears and as mentioned in Section 1.1 a light wave is created that extends in all directions, in the glass and in the free space behind the glass surface. In other words light rays are going out in all directions from the actual atom.

If now a light ray hits a surface atom (Fig. 1.3), a second ray can be found with a travelling way difference of one wavelength that excites another surface atom in a certain distance. Both atoms are forced to oscillate as explained above reaching their maximum surface charges and field strength at the same time and thus emitting light waves with coinciding maxima according to Fig. 1.3. The same two atoms emit also waves with coinciding maxima in one distinct direction, namely with the same angle of incidence (between the direction perpendicular to the surface and the incident ray), but with inverse sign, that means a beam is generated that leaves the surface into the bulk of the glass and is symmetric to the incident beam (*reflection*).

The two atoms under consideration also emit waves in the space outside the glass that would in principle continue the incident rays.

Nevertheless, the speed of light is higher in the free space than in the glass since no time consuming charging actions as above are necessary. So the wavelength is larger than in the glass, what means that the maxima of the two rays regarded here coincide only, if the rays leave the glass under an angle of incidence larger than the angle of incidence in the glass (*refraction*).

If the difference between the wavelength in the glass and outside is too large, the direction of the refracted wave is in parallel to the surface omitting any light leaving the glass and *total reflection* appears, being the reason for light conduction through glass fibers without essential losses to outside.

It should be mentioned that in theoretical optics the phenomena described above are explained only formally by solving the so called Maxwell equations [3] that relate the temporal and spatial variations of the electrical field strength and the associated magnetic fields to each other under the condition of a light wave incident on a surface separating two media with different materials and different speed of light. This method gives of course no insight into the mechanism of the phenomena treated here, but allows to get numerical results for the *refraction index* (speed of light in vacuum divided by that of glass), being of crucial importance for many optical devices, as microscopes and telescopes.

1.4 Principal Mechanism of Light Amplification by Stimulated Emission of Radiation

Lasers are devices that generate a directed beam of light based on the emission of light by atoms stimulated by an incident light wave thus loosing energy (*stimulated emission*). The latter process is the reverse phenomenon to light absorption by atoms that gain energy and must exist to allow equilibrium of atoms with low energy and those with high energy. Since the latter are rare compared to those with low energy, absorption usually prevails stimulated emission. Nevertheless if energy either kind is supplied to the atoms (*pumping*) the high energy atoms may dominate and then an incident light wave is amplified, although only if the light wave has a frequency or wavelength matched to the atoms (*resonance*).

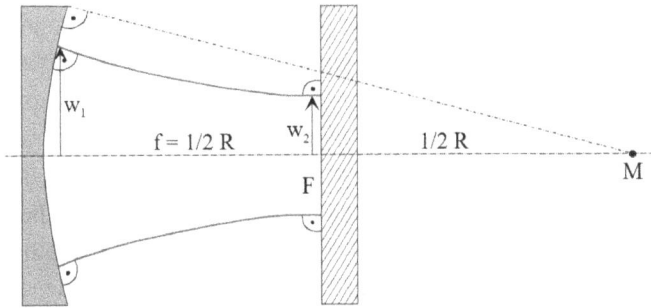

Fig. 1.4 Optical resonator consisting of two mirrors, one spherical and completely reflecting and one plane and partly transmissive. The space between the mirrors is filled with active medium that amplifies light due to pumping. The plane mirror extracts a part of the internal beam for practical use.

Source: D. Schuöcker, Spanlose Fertigung, Fig. 3.2.3 Oldenbourg, München 2004.

Pumping is preferably performed either by electrical energy (diode lasers, CO_2 laser) or by light (solid state lasers), all treated below.

So far an amplifier for light is available that can be made a light generator if two mirrors are arranged at the two ends of the amplifier (Fig. 1.4) serving for a feed back of amplified radiation, thus building up a wave with permanently increasing strength (electrical field strength of the light wave) between the mirrors. The strength of the radiation rises from one round trip to the next as long as the amplification along the round trip is stronger than the losses especially at the mirrors. The latter losses are caused mainly by transmission at one of the two mirrors that is partly transmissive to allow the extraction of a part of the generated light. Moreover, both mirrors cut off a part of the radial intensity distribution of the internal beam due to their finite extension causing further losses. Eventually the permanently rising amplification equals the losses and then an equilibrium state with constant strength of the generated light is obtained.

To enable feedback and amplification the internal beam must be reproduced after a full round trip what can only be achieved if the beam hits both mirrors perpendicularly as being obvious from Fig. 1.4. A further condition for full reproduction is that the travelling way-twice the distance between the mirrors is a multiple of the wavelength, a condition that ensures that the electrical field strength of the internal light wave is zero at both mirrors. The latter phenomenon is caused by charge displacements compensating the component of the field strength that is in parallel to the mirrors (the latter phenomenon is treated in detail in Section 1.7). Therefore possible wavelengths are given by the distance of the two mirrors that are thus called *optical resonator*. Since the wavelength range for amplification of the active medium is usually relatively wide, more than one wavelength can be generated. The respective beams are called *longitudinal modes*.

Since only waves that propagate perpendicularly with respect to the plane mirror can remain in the resonator, the light coming out of the laser propagates only in one distinct direction, thus allowing sharp focusing to high energy densities well suited for material processing for instance by melting and evaporating metals.

Resonators serve not only for feedback and directed output of the beam, but also ensure that the lateral light losses as mentioned above remain weak, a task that is achieved by using one spherical mirror that focuses the beam inside the resonator on the second plane mirror. The latter also defines the radius of the beam outside of the laser becoming obvious from Fig. 1.4.

If the radius of the spherical mirror R is small and the length of the resonator L is big, the beam will diverge strongly between the mirrors and the mirrors will cut off a considerable part of the beam cross section what means that only the main lobe of the beam is surviving during feedback and no side lobes can be amplified yielding a beam similar to a Gaussian beam with high quality and high focusability. The exact analysis of an optical resonator, that is unavoidable if dimensions must be calculated, shows that the expression $R^2/\lambda L$, the *Fresnel number*, must be one or smaller to ensure the exclusive generation of a Gaussian beam.

As already mentioned, the output power of a laser is determined by light gain due to stimulated emission and light losses at the edges of the mirrors, imperfect reflection by the latter and output. Of course, amplification depends on pumping strength. Nowadays 100 kW can be reached by CO_2 and solid state lasers, although with rising power the mirrors become larger to avoid damage by overheating and in consequence the beam parameter product and thus the beam quality becomes worse [4].

1.5 Focusing a Gaussian Beam

In order to obtain strongest heating as most useful for the majority of laser manufacturing processes, the beam power must be high and the focus small, what means that the intensity at the workpiece becomes high as it can be obtained by focusing the beam either by lenses or mirrors.

If a lens with radius D/2 focuses a Gaussian beam, the beam radius reduces permanently until a minimum radius w_0 is reached at a distance f (focal length) followed by a continuously diverging radius, where the angle of beam constriction and that of the diverging beam are equal due to the symmetry with respect to the focal plane. The latter angle, the divergence, is approximately equal to the lens radius D/2 divided by the focal length f. Due to the beam parameter product (minimum beam diameter times divergence) given by the wavelength λ and π, by substituting the divergence the minimum beam radius in the focus becomes $w_0 = 2 \lambda f/\pi D$.

1.6 Nonlinear Optics

If a light wave is incident on a nonmetallic material, every atom hit by the wave is polarized by its electrical field, that means electrons are drawn to one part of the atom surface and the opposite surface appears to be positively charged. These charges compensate the incident field totally but act also as sources of a light wave leaving the atom. As long as the electrical field is relatively weak, the strength of the surface charges of the atom is strictly proportional to the field strength of the wave.

Nevertheless if the field becomes very strong, the atoms have not enough electrons to compensate the field of the incident wave and the light wave emitted by the atom is no longer proportional to that of the incident wave. Due to a lack of electrons the field strength caused by the displaced electrons remains constant until the incident field reaches again a lower strength, which means that the wave form of the emitted wave is no longer purely sinusoidal, but consists of a fundamental wave with the frequency of the incident light superimposed by waves with a multiple of the basic frequency, especially with the doubled frequency. Practically *Lithiumniobat (LiNb)* shows such a nonlinearity to a high extent. If a Nd:YAG Laser is used as the source ($\lambda = 1{,}064$ μm) and a LiNb crystal is irradiated, green light ($\lambda = 0{,}532$ μm) is obtained, where a light power of some Watts can be reached [5, 6].

1.7 Absorption of Light by Metals

Absorption of light by solids depends on various parameters, as material, wavelength, polarization (electrical field has always only one distinct direction) and angle of incidence (angle between the direction perpendicular to the absorbing surface and the direction of propagation of the light wave).

Two polarization directions must be distinguished: The electrical field is in parallel to the absorbing surface s polarization) or it is in the plane of incidence given by the incident beam and the direction perpendicular to the absorbing surface (p polarization) (see Fig. 1.5).

In the first case of s polarization and also for p polarization at an angle of incidence equal to zero, the electrical field is in parallel to the surface and absorption is low due to the compensation of the field by charges on the surface that are induced by the field of the incident wave displacing electrons and leaving positive surface charges. The latter give rise to an electrical field of opposite direction thus compensating the incident field and preventing the incident wave to enter the solid what means very low absorption (Fig. 1.6).

If the angle of incidence at p polarization rises, a component of the field perpendicular to the surface appears and becomes stronger and stronger reaching a maximum value at striping incidence.

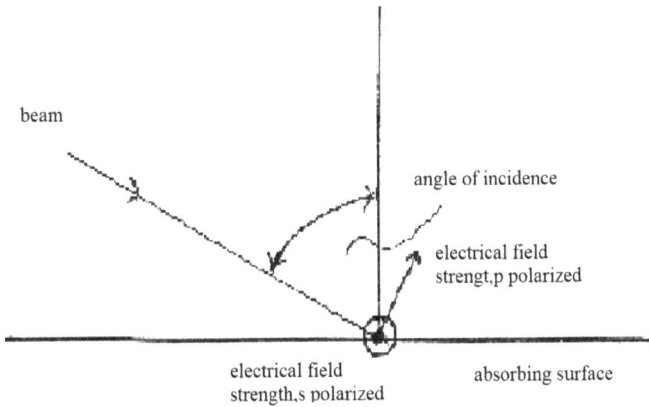

beam

angle of incidence

electrical field
strengt,p polarized

electrical field
strength,s polarized

absorbing surface

Fig. 1.5 Absorption for p and s polarization.

Source: Authors.

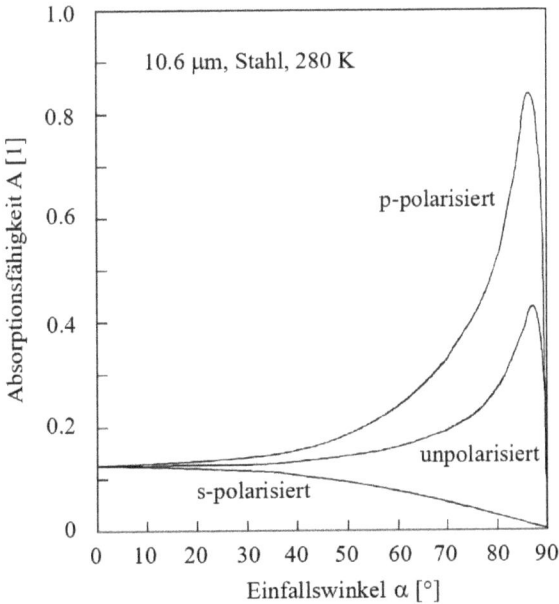

10.6 μm, Stahl, 280 K

p-polarisiert

unpolarisiert

s-polarisiert

Fig. 1.6 Absorption dependent on polarization and angle of incidence.

Source: D. Schuöcker, Spanlose Fertigung, Fig. 3.4.3 Oldenbourg, München 2004.

The latter perpendicular component of the field of the incident wave leads now to absorption due to the following mechanism:

The incident light wave causes the electrons to oscillate forth and back. Thus the electrons and the positively charged atoms fixed in the metal form oscillating dipoles that emit themselves a light wave in a wide range of directions.

If the electrical conductivity is high, many free electrons are available and the incident light wave is absorbed totally in a thin layer at the surface and the wave emitted by the above dipoles leaves soon the metal and reaches the free space above the surface acting as reflected radiation and the incident radiation does not penetrate considerably into the solid but is strongly reflected.

If the solid is a poor conductor not so many electrons are available to absorb the incident wave and so the radiation penetrates much deeper into the metal. The radiation emitted by the oscillating dipoles also finds space up to the surface to be absorbed and therefore absorption of the incident wave is strong and not much radiation reaches the surface and thus reflection is weak [7].

If the solid is an insulator, practically no free electrons are available and the electrical field of the incident light wave can only act on the lattice atoms, for lower frequencies as in the case of a CO_2 laser by polarizing the atom and then moving it due to its polarization charges thus consuming energy from the wave and absorbing the latter strongly.

For higher frequencies as of the Nd:YAG laser, inertia obstructs a movement of the atoms and thus weak absorption results (Fig. 1.7).

Somewhat different are things if green light with its relatively high frequency or near ultraviolet light as generated by excimer lasers is absorbed to a larger amount, since then the energy that is exchanged with the atoms, the light quantum, is high enough to allow the excitation of higher energy levels of the atomic electrons absorbed by the atoms. The latter phenomenon is of special importance for highly conducting materials as gold, copper, etc [8].

LASER

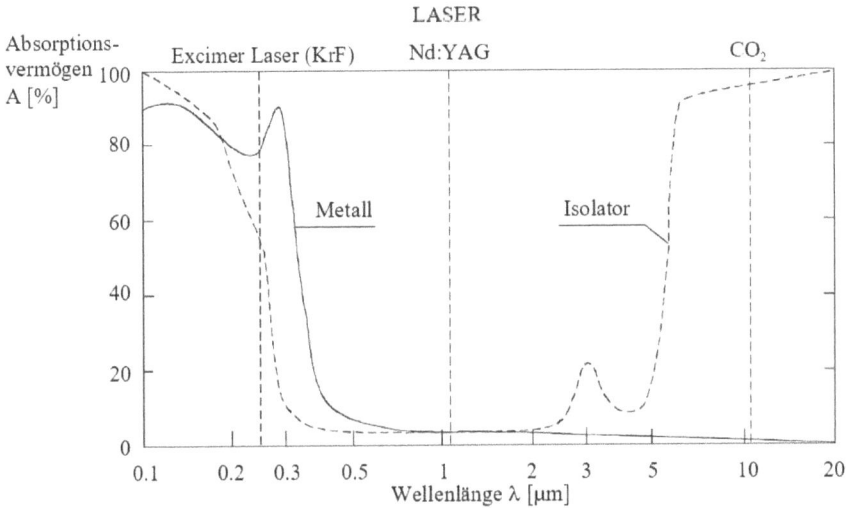

Fig. 1.7 Experimental values for the absorption dependent on material and wavelength for metal and insulator at the wavelength of Nd:YAG and of CO_2 laser for perpendicular incidence.

Source: D. Schuöcker, High power lasers in production engineering, WSP and Imperial College Press, Singapur/London 1999.

1.8 Heating of Workpiece by a Focused Laser Beam

In the case of a relative movement between laser beam and workpiece, the laser beam constantly moves from already heated zones of the workpiece to cold zones that have not yet been heated, which have to be heated by the laser beam first.

This allows a balance between the supplied laser energy and the energy needed to heat cold regions to be achieved at an appropriate speed and yields a temperature distribution with a maximum in the centre of the focus of the laser radiation on the workpiece surface and a drop in temperature in all directions on the surface and into the depth of the workpiece.

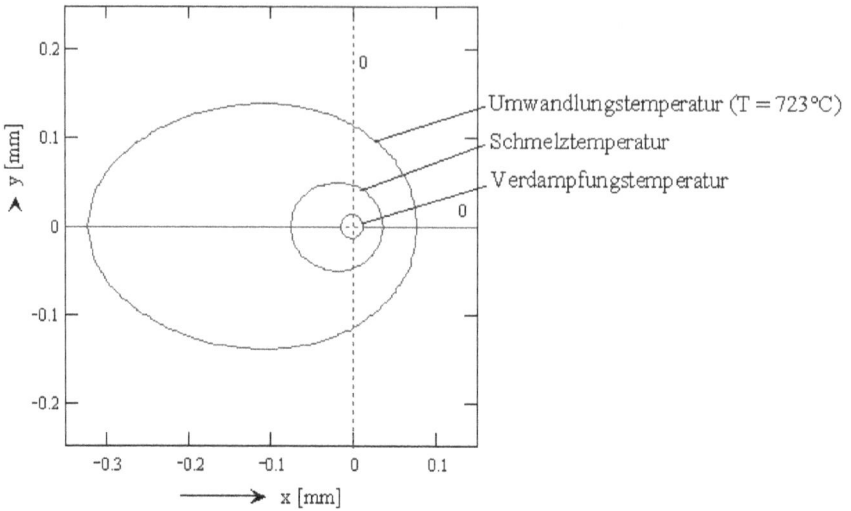

Fig. 1.8   Isotherms of a moving point source, steel (d = 5 mm, A.PL = 1,000 W, v = 5 m/min).

Source: D.Schuöcker, High power lasers in production engineering, WSP and Imperial College Press, Singapur/London 1999.

Figure 1.8 shows calculated values of this temperature distribution for steel with a thickness of 5 mm and an absorbed laser power of 1,000 W and a point like focus and a travel speed of 5 m/min.

It can be seen that the isotherms are, for example for the melting temperature T_m in the direction of the movement of the laser beam over the workpiece surface relatively close to the focus, which is caused by the fact that due to the already mentioned penetration of the temperature distribution into still cold regions a strong heat supply is necessary and therefore the gradient of the temperature must be large, which leads to the isotherms being close together. Incidentally, the opposite is true at the back of the isotherms. A mean value for the distance of the isotherms is therefore obtained in the two lateral directions.

If one now takes into account the finite expansion of the focal spot with the radius r_F, one can again mathematically obtain curves for the maximum temperature.

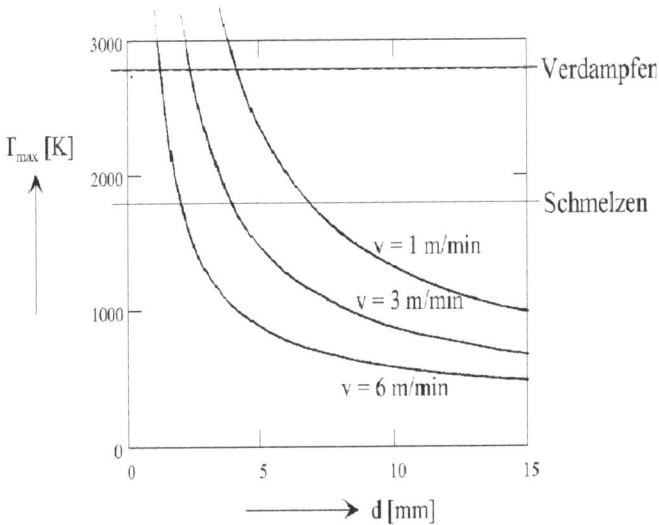

Fig. 1.9   Maximum temperature in the focus of the laser beam.

Source: D. Schuöcker, High power lasers in production engineering, WSP and Imperial College Press, Singapur/London 1999.

Figure 1.9 then shows the dependence of the maximum temperature reached at the workpiece surface in steel for various laser outputs and workpiece thicknesses, from which, for example, it can be extracted immediately, up to what thickness with given laser power laser cutting, i.e., melting is possible at all.

The diagram shows that with increasing relative speed between the laser beam and the workpiece, the temperatures reached in the workpiece decreases, which is due to the fact that at higher speeds the cooling effect is amplified by heating up unheated regions.

Furthermore it can also be concluded that in order to reach a certain temperature in the workpiece, such as the melting temperature required for laser cutting, the product from speed and workpiece thickness grows with increasing laser power, a relation that basically applies to all laser processing processes, because these differ essentially only by the necessary temperature.

References

1. F. Graham Smith, Terry A. King, Dan Wilkins: Optics and Photonics: An Introduction. John Wiley & Sons, 2007, p.240 f. ISBN 978-0-470-01783-8.
2. Gerd Herziger, Horst Weber: Der Laser — Grundlagen und Anwendungen. Physik-Verlag, Weinheim 1972.
3. John David Jackson, Classical Electrodynamics. John Wiley, New York NY 1962 (3. edition.1999) ISBN 0-471-30932-X.
4. J. W. Goodman, Fourier Optics, Wiley.
5. P. N. Butcher, D. Cotter: The Elements of Nonlinear Optics; Wiley 1984.
6. Encyclopedia of laser physics and technology. Retrieved 2006-11-04., Frequency doubling.
7. Wikipedia-the free encyclopedia: Fresnel equations, 2021.
8. Helmut Hügel, Strahlwerkzeug Laser — Eine Einführung, Springer Vieweg, 1992.

Chapter 2

Laser Equipment

2.1 Carbon Dioxide Laser

2.1.1 *Gaseous plasma*

In a gas the atoms consisting of negatively charged electrons and positively charged ions are initially electrically neutral. Nevertheless due to always present energetic radiation some electrons are deliberated from the atoms since the latter absorb radiation and gain thus the energy necessary to overcome the attraction by the positive ions (*ionization*). Thus, free electrons are present and can move around freely. If now two metal plates, the electrodes, are arranged in the gas and a high-voltage is applied (Fig. 2.1) the electrons are attracted by the positive electrode and gain thus high energy sufficient to deliberate further electrons from the atoms due

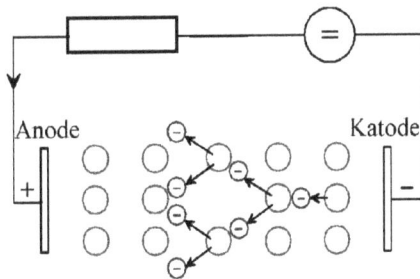

Fig. 2.1 Formation of a plasma.

Source: D. Schuöcker, Spanlose Fertigung, Fig. 1.4.1 Oldenbourg, München 2004.

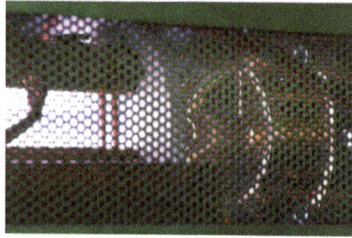

Fig. 2.2   Pinkish blue luminosity of the plasma in a CO_2 laser.
Source: D. Schuöcker, Spanlose Fertigung, Fig. 3.3.9 Oldenbourg, München 2004.

to collisions. Quite similar as inertia must be overcome by lifting a certain load into height, in order to deliberate an electron from an atom the attraction by the positive nucleus must be overcome by energy supply. After a short time many free electrons and positively charged atoms are present in the gas that is now able to conduct electrical currents. In this state the gas is called a *plasma* and shows a certain glow due to electrons recombining with atoms that are emitting the ionization energy mentioned above as light (see Fig. 2.2).

2.1.2 *Light amplification mechanism*

In lasers in which the light-amplifying medium is a gas such as carbon dioxide, the latter is ionized by the supply of electrical energy as mentioned above and a plasma is formed, which means that there are many freely moving electrons. The latter are strongly accelerated by the applied voltage in the range of a few 1,000 V and then hit the carbon dioxide molecules and cause vibrations of the atoms in the molecule (Fig. 2.3).

The CO_2 molecule consists of one carbon and two adjacent oxygen atoms, that are all fixed together by elastic forces and can, thus carry out oscillations around their equilibrium positions. In the two most important vibration states, the carbon atom rests and the two oxygen atoms move always in the same direction (asymmetric vibration mode). In the second state the two oxygen atoms move in opposite directions (symmetric mode). The first state has a higher kinetic energy since the center of gravity moves, what is associated to kinetic energy that adds to the potential

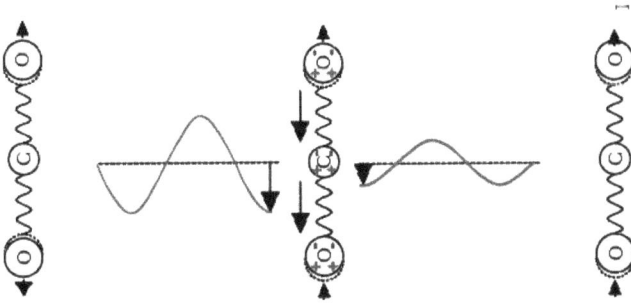

Fig. 2.3 Vibrations and transitions of the CO_2 molecule.

Source: D. Schuöcker, High power lasers in production engineering, Fig. 4.7, WSP and Imperial College Press, Singapur/London 1999.

energy of the atoms. In the second case the energy content is thus lower. The three atoms are polarized by the electrical field strength of the wave and opposite charges appears on opposite surfaces of the atoms (see Fig. 2.3). If the frequency of the light hitting the atoms is equal to the frequency of the vibrations and the molecule has been excited to the asymmetric vibration due to hitting electrons after some time it is forced to change to the symmetric vibration mode in order to match to the symmetry of the surface charges. After this transition the molecule energy is lower as mentioned above and this can only be performed by emission of light, that means simulated emission with a wavelength of 10.6 μm (far IR) takes place.

2.1.3 *Layout of CO_2 lasers*

A conventional CO_2 laser consists now in principle of a tube filled with CO_2 and some other augmenting gases as Nitrogen and Helium. Two electrodes on both ends of the tube allow the supply of electrical energy to form a plasma and to charge the CO_2 molecules with excitation energy. The setup is completed by two mirrors providing feedback, direction selection and extraction of the laser beam. One of the mirrors, the output window must be semitransparent what can only be achieved in the far IR wavelength range of the CO_2 laser with a semiconductor, especially with ZnSe.

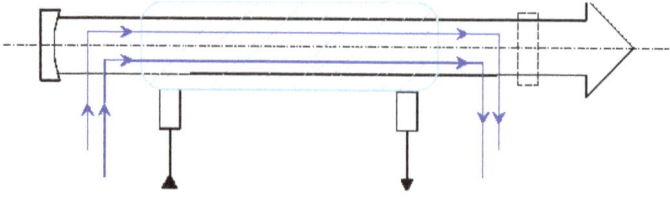

Fig. 2.4 Layout of a carbon dioxide laser.

Source: D. Schuöcker, High power lasers in production engineering, Fig. 4.12, WSP and Imperial College Press, Singapur/London 1999.

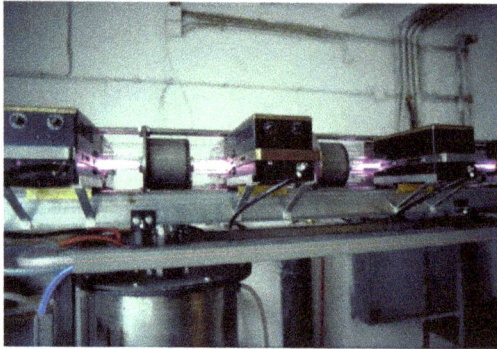

Fig. 2.5 CO$_2$ Laser with RF energy supply and fast gas flow, beam power 2 kW, developed at TU Vienna.

Source: D. Schuöcker, Spanlose Fertigung, Fig. 3.3.7, Oldenbourg, München 2004.

Since the efficiency of the conversion of electrical energy to light is imperfect, the laser gases are heated and thus the gas must be circulated to a water cooled heat exchanger. A laser of that kind is shown by Fig. 2.4.

A practical construction of such a laser is shown in Fig. 2.5. The electrical energy is supplied by RF currents in order to be able to put more energy into the unit volume and the laser has eight plasma columns to reach a beam power of 2 kW.

Strong energy supply by DC currents leads to overheating of the electrodes due to the impact of charged particles, thus limiting the power per volume capability. Nevertheless change to RF currents that are supplied not directly by the electrodes to the plasma but by charging and

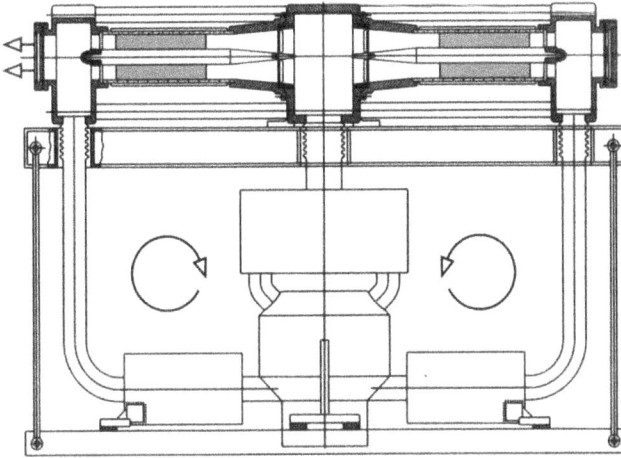

Fig. 2.6 Schematic layout of a coaxial laser.

Source: Authors.

discharging the capacitor formed by the electrodes that may even be fixed outside the plasma confining glass tube, allows enhanced power supply.

A design capable of very high beam power up to 100 kW, the *coaxial laser* (Fig. 2.6) uses a hollow cylindrical outer electrodes and a cylindrical inner electrode to build up a hollow cylindrical $CO_2/N_2/He$ plasma. With a toroidal, completely reflecting metal mirror and a semitransparent ZnSe output window a hollow cylindrical beam is generated, that can be well focused since theoretical considerations show, that focusing a light beam results in a Fourier transformation of the beam cross section, what means that the ring shaped beam cross section transforms to a circular spot [1]. Gas cooling is carried out by a strong turbo blower and three heat exchangers. Figure 2.7 shows the practical laser with glowing plasma and Fig. 2.8 a stone with a glowing region made with the unfocused beam with a light power of 6 kW.

A laser of the latter kind has been invented at TU Vienna by one of the authors and developed successfully by Dr. Markus Bohrer from the same institute. A leading laser company adopted the principle of the coaxial laser and offers it now on the market.

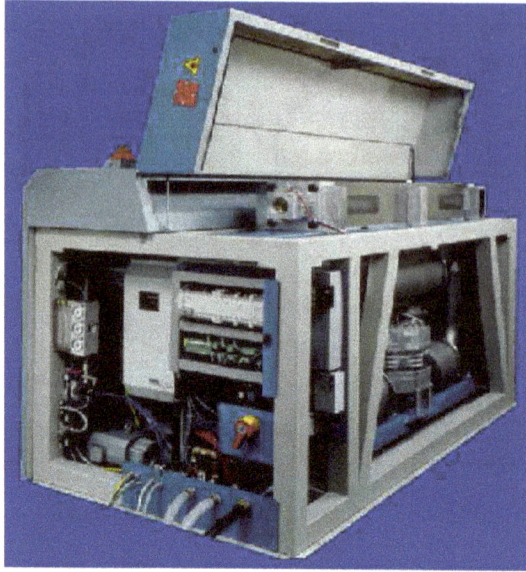

Fig. 2.7   Coaxial laser 6 kW beam power.

Source: D. Schuöcker, Spanlose Fertigung, Fig. 3.3.10 Oldenbourg, München 2004.

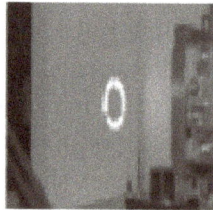

Fig. 2.8   Burn-in of a 6 kW beam of the coaxial laser on a stone.

Source: Authors.

2.2 Diode Lasers

2.2.1 *Introductory remarks*

Diode lasers have caused a revolution in laser source technique quite similar to that caused by 3D printing in production technology, since they generate in a very small volume high beam power, although it is not very well focused. The latter disadvantage can be overcome by using these lasers not directly for material processing but for pumping of solid-state lasers as most recent disc lasers and fiber lasers.

The beam power generated by these Lasers is in the order of magnitude of a maximum of some Watt and thus far away from the high power levels that are necessary for material processing, a problem that can be solved by combining a large number of these devices to one single device with power up to a maximum of nearly 20 kW.

2.2.2 *Light generation in semiconductors*

Laser diodes use two kinds of semiconductors for the generation of light, namely so-called in *n*-semiconductors that contain electrons that are deliberated by impurities (*donors*) and not bound to atoms and can thus move freely around and transport negative charge and are able to conduct electrical current. Quite similar so-called p-semiconductors contain a lot of atoms having lost one electron to other impurities (*acceptors*) and are therefore positively charged and named *holes*. They can also move around freely due to hopping of the missing electron from one atom to the other and can transport positive charge and conduct thus also electrical currents.

If these two kinds of semiconductors touch each other along a certain cross section and a voltage is applied to the two parts of the arrangement (Fig. 2.9) electrons move to the positive pole and holes to the negative one, both meeting in a narrow zone around the cross section mentioned above where initially electrons have been lost by diffusion to the p-side and vice versa holes to the n-side, therefore called *depletion zone*. Since electrons and holes attract each other due to their different charges they recombine to an electrically neutral atom, where the energy that had to be used to liberate the electron from its atom, the so-called *ionization energy* is set free as a light quantum, called *photon*. So far the device that resembles a part of transistor forms a light emitting diode (LED).

transition region

n semiconductor

p semiconductor

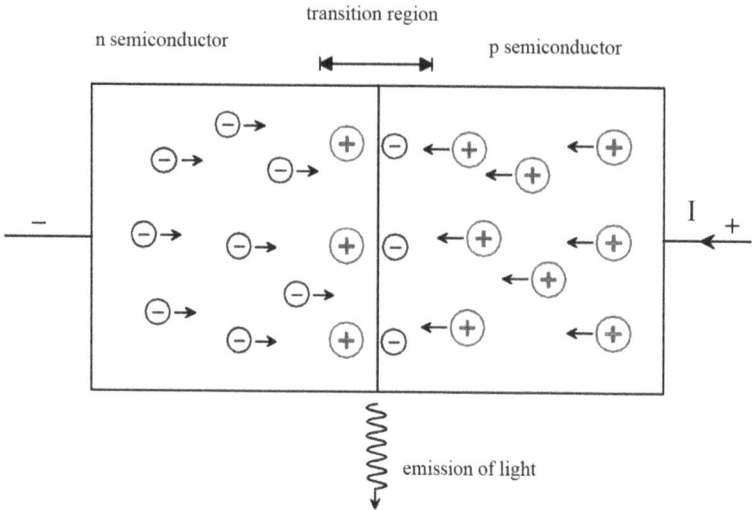

I

emission of light

Fig. 2.9   Light emission in a current leading pn-junction.

Source: D. Schuöcker, Spanlose Fertigung, Fig. 3.3.11 Oldenbourg, München 2004.

2.2.3 *Light amplification in semiconductors*

So far the devices mentioned in Section 2.2.2 do not generate a laser beam since the emitted light propagates in all directions. To obtain a directed beam of light it is necessary to use the property of the pn-junction not only to emit light by recombination but also to be able to amplify light. The latter phenomenon can be understood, since the deliberation of an electron from its atom due to the absorption of light must also be associated to the reverse action where light hitting the electron causes it to recombine with a hole, thus amplifying the incident light wave (*stimulated emission*). By the way the latter reciprocity is necessary to allow an equilibrium of charged and uncharged particles. Clearly absorption and stimulated emission can only take place if the energy that is supplied by a light wave to the atom is equal to the ionization energy because if it is too low it would be insufficient or too much energy would lead to the problem how to dissipate the excess energy.

A large amount of stimulated emissions compared to absorptions depends on the number of free electrons and holes and thus on an

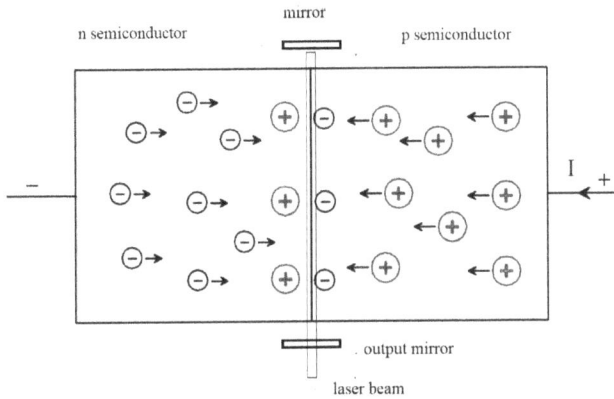

Fig. 2.10 Diode laser.

Source: Authors.

especially high number of donors and acceptors that generate free electrons in the n part of the device and holes in the other part (*degenerate semiconductor*). Moreover a high flow of current transports many electrons and holes in the depletion zone what also enhances stimulated emission and is thus necessary for laser function. In that way the depletion zone becomes light amplifying and by adding two mirrors on both sides of the latter, that perform feedback of the amplified light, selecting only waves travelling in a direction perpendicular to the mirrors establish a laser with intense light propagating in one direction (see Fig. 2.10).

2.2.4 *Setup of semiconductor lasers*

Practically laser diodes (Fig. 2.11) are built by epitaxial growth of various thin layers with a length of 500 μm, a width of 10 μm as well as a thickness of 50 μm made from GaAs with different impurities for use as n or p semiconductors.

The lowest layer is the metallic contact for the n-semiconductor followed by a thick layer of n-semiconductor material, followed by only some micrometer wide and thick layer that enforces a high current density in the active zone. Above that zone a much thicker layer of

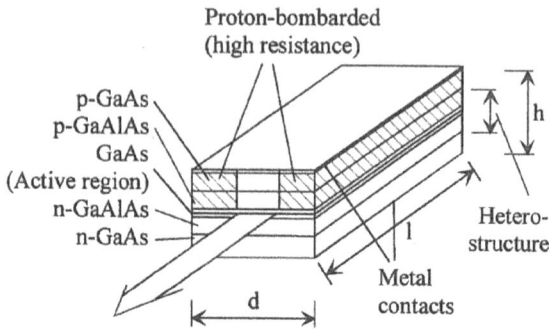

Fig. 2.11 Construction scheme of a laser diode (Wikipedia).

Source: By Photon, CC BY-SA 3.0, https://commons.wikimedia.org/w/index.php?curid=834343.

p-semiconductor is arranged finally followed by a metallic contact. The necessary mirrors are formed by the free surfaces of the device due to its changing refraction index that leads to reflection. With a current of a few Amps the laser generates power up to several Watts, the efficiency can be up to 70% preferably in the near infrared. Disadvantageously the laser creates an elliptical beam what makes focusing difficult [2].

2.2.5 Semiconductor lasers for multikilowatt operation

High laser power sufficient for material processing can be obtained if a large number of single laser diodes, practically up to 25, are arranged on the edge of a thin substrate where a light power of 200 W can be generated as a line of light (Fig. 2.12).

The substrate contains tiny little channels for the cooling water supplied to the individual diodes. These micro channels begin and end at two central holes that supply the main coolant flow. Such an arrangement is called a *laser diode bar*. Stapling these bars forms a *stack* that allows to obtain Kilowatt light power, where all the holes mentioned above form tubes for the cooling water. The beam generated by this arrangement has a large rectangular cross section. Because it consists of hundreds of individual beams it must be focused by mirrors and lenses as far as possible finally reaching a focus of 1 mm².

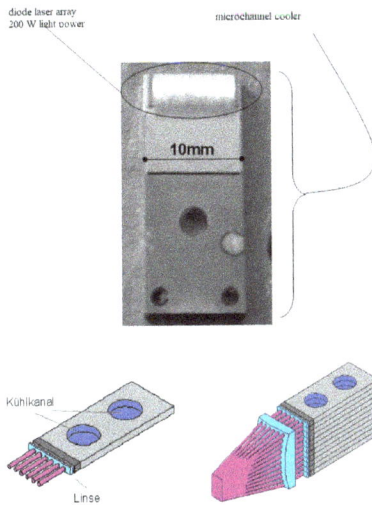

Fig. 2.12   Laser bar on a substrate with cooling channels (top and bottom-left) and a stack of laser bars with focusing lens (bottom right).

Top image source. Authors.
Bottom images: D. Schuöcker, Spanlose Fertigung, Fig. 3.3.12, Oldenbourg, München 2004.

2.3 Solid State Lasers

2.3.1 *Rod laser*

Solid state lasers use as light amplifying medium a host crystal that is transmissive for visible and infrared near as YAG (aluminum yttrium garnet), shaped as a cylindrical rod. The latter contains atoms that serve to amplify light by stimulated emission if irradiated with visible or near IR. Example is Ytterbium that amplified light with a wavelength of 1.0 μm. Since the laser has an imperfect efficiency, the host is heated and cannot dissipate the heat to an appropriate amount and heats therefore thus suffering from deformations that cause distortions of a Gaussian beam due to a geometry that deviates from rotational symmetry. Therefore a pure Gaussian beam with rotational symmetry is spoiled with additional modes reducing beam quality. Nevertheless new construction schemes avoid the latter shortcoming by using an amplifying material, either a thin little disk from YAG material or a glass fiber enriched with Ytterbium atoms as active players (disk and fiber lasers). The latter devices are treated in the following [2].

2.3.2 *Disk laser*

The rod laser could be improved by the so-called disc laser (Fig. 2.13), in which the radiation of several diodes lasers is focused on a thin, disc-shaped sheet of Yb doped YAG material, thus achieving amplification. The sheet is mounted on a water-cooled heat sink and also mirrored at its back and thus represents a part of the resonator, which is completed by a second partially permeable mirror, which allows the decoupling of a part of the radiation generated in the resonator. Due to the good heat dissipation of platelets to the heat sink, the disadvantages of the classic solid-state laser as a deformation of the amplifying material and thus beam degradation are avoided. Moreover the geometry with rather small mirrors at a large distance leads to the preferred generation of a Gaussian beam and thus to a very low beam parameter product in the order of magnitude of 1 mm·mrad, compared to that of a rod laser with 10 mm·mrad. The consequence is a very low divergence, that allows to use the beam for processing of remote workpieces, as in remote laser welding. Another advantage is the good focusability that is very well suited for the feeding of glass fibers, that can be used with solid state lasers due to their wavelength in the near infrared, a radiation where glass is transmissive.

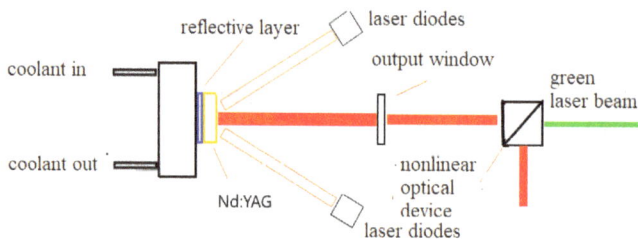

Fig. 2.13   Schematic layout of a disk laser with frequency doubling.

Source: D. Schüöcker, 3D Druck aus Kupfer, Gold etc, METALL Monatsschrift des Wirtschaftsverlages Wien, p. 30, 7–8, 2020.

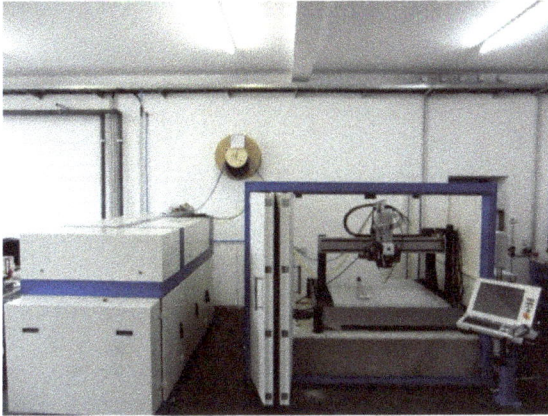

Fig. 2.14 (Left) 6 kW disk laser, (right) gantry robot system with scanner, (middle) fiber connecting laser and robot, later also connected to a welding robot and a 3D print facility (Laser Center Gmunden 2013).

Source: Authors and coworkers.

Lasers of the kind are offered for CW beam power up to the multi 10 kW range and show efficiencies for the conversion of electrical energy in IR light up to multi 10%.

Advantageously most disk lasers for materials processing are equipped with several connectors for glass fibers, what makes them very well suited as central light source for a multitude of processing station (Fig. 2.14).

2.3.3 *Fiber laser*

Disc laser transform the poorly focusable radiation from laser diodes into a well-focused laser beam. The same can be done with another geometry, that of the so-called fiber laser (see Fig. 2.15): such lasers use as an active medium a thin glass fiber, as it is used for the transmission of light. The latter is doped with the atoms of rare earths such as erbium, neodymium or ytterbium. Light amplification is achieved by the fact that the radiation from numerous laser diodes is coupled by glass fibers to the above mentioned active fiber and there supply the mentioned atoms energy, which then emits them as light. At both ends of the active fiber, as with each

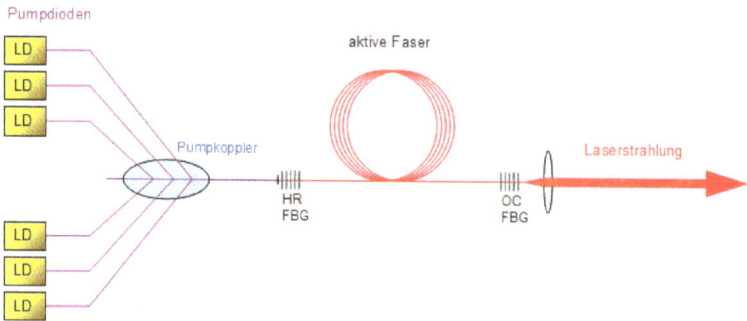

Fig. 2.15 Setup of a multi kW fiber laser.

Source: (Von MyNick — Eigenes Werk, CC BY-SA 3.0, https://commons.wikimedia.org/w/index. php?curid=11800989).

laser, two *fiber Bragg gratings* [3] acting as mirrors for the laser radiation are attached, one 100 percent reflective and the other partially transparent. The latter provide the necessary feedback and the decoupling of the amplified light beam, which turns the light amplifier into a laser source. Since the fiber, which is very long in order to obtain strong light amplification and thus light high output power, has a large surface, it can be cooled well to dissipate unavoidable energy losses. This has already enabled light outputs of up to 100 kW to be generated. The main advantage of these lasers is the extremely good focusability, which makes it possible to achieve high light power densities, just as they are well suited for cutting, ablation and deep penetration welding. This type of laser was developed by Russian scientists and tested relatively early at the Vienna University of Technology together with the University of Ljubljana/ Slovenia, which illustrates Fig. 2.16.

A variant of this design uses only a single high-performance laser diode, which is directly coupled into one end of the light-enhancing glass fiber via a focusing optics, but a solution which is more likely to be used for small outputs and was analyzed mathematically at the above-mentioned institute many years ago, thus proving its feasibility [5].

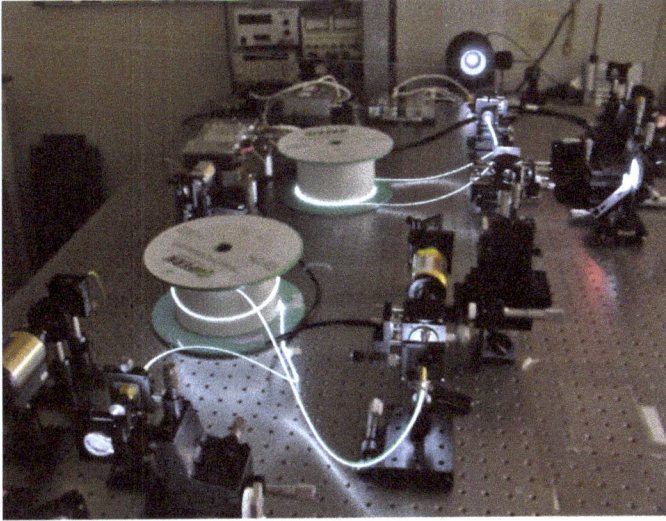

Fig. 2.16 Experimental setup of a low-power fiber laser (TU Wien).

Source: Authors and coworkers.

2.4 Pulsed Lasers and Femtosecond Laser

Some laser types can emit light both continuously and in pulsed form. Examples are the most commonly used devices such as diode lasers, solid-state lasers and CO_2 lasers. However, a smaller group of lasers can only be operated in pulsed mode, such as atmospheric pressure CO_2 lasers or excimer lasers that emit ultraviolet light through the decay of unstable noble gas molecules and finally also femtosecond lasers.

Pulsed laser radiation can be used to improve the quality of laser processing of workpieces with complex geometry or of high-melting materials and for processing without heat penetrating into the depth of the workpiece.

Normal pulsed lasers like CO_2 lasers, where the pulse peak power is equal to the power in the continuous operation mode, is used in laser cutting of sharp bends that are passed with a lower speed that asks for

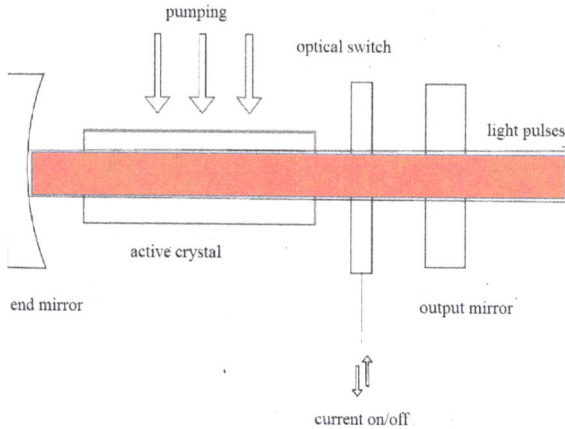

Fig. 2.17   Q switched laser with super pulsing.

Source: D. Schuöcker, Gepulsted Licht etc, METALL Monatsschrift des Wirtschaftsverlages Wien, p. 24, 1–2 2020.

reduced laser power to avoid overheating of the material. Pulsing reduces the average power supplied to the workpiece, but in the pulse peaks supplies enough energy to achieve melting, which is necessary for laser cutting.

Super pulsed lasers generate pulses that are far higher than in continuous operation what is achieved by so-called *Q switching*, that alters quality of the resonator (see Fig. 2.17). This is achieved by an optical switch placed between one of the mirrors and the light amplifying medium. The latter is initially closed, so no light can propagate in the resonator and more and more atoms are excited to the upper laser energy level. If then suddenly the optical switch is opened, the laser begins to emit light where the stored energy is deliberated in a very high pulse. These pulses that are much higher than the continuous output power are used to cut materials difficult to melt due to a high melting point as, e.g., chromium oxide as formed during cutting of stainless steels.

Femtosecond lasers are particularly interesting because of their extraordinarily short but very high pulses (pulse duration 10^{-15} s). The latter use the phenomenon that usually the active media amplify light in a wide range of wavelengths and therefore light with a multitude of wavelengths (*longitudinal modes*) is generated. These modes interact with each other like interfering waves at the sea and create a very high peak in the resonator similar to giant waves on the sea that cause tsunamis. This type of laser can be used for treating workpieces without any heating of the bulk the workpiece. Femtosecond laser can not only be used to perform machining operations in metalworking but are also very successful in other applications as, e.g., for surgery, especially of the eyes.

Practically the *Titanium Sapphire Laser* can be used to generate femtopulses. The latter laser has an extremely wide wavelength range of amplification and thus if pumped, e.g., with a Nd:YAG laser, light with many wavelengths (around green light) as mentioned above is generated. All these longitudinal modes superimpose in a certain location to a huge pulse, quite similar as Fourier analysis shows that a narrow and high pulse is obtained by the synthesis of harmonic functions with a large number of wavelengths. The light pulse mentioned above changes its position due to instabilities of the modes and runs thus forth and back between the mirrors with further enhancement. In that way pulse lengths as small as 5 fs (10^{-15} s) and beam power of Terawatt (10^{12} W) can bereached [4].

2.5 Light Conducting Fibers

Ordinary silicon glass is transparent for the radiation of diode and solid state lasers and can thus be used to transmit laser beams. Nevertheless the beam spreads up in the glass due to its divergence what asks for a rising extension of the glass body with rising distance from the source. The latter problem can be avoided if the glass body has the shape of a narrow fiber since the surface of the latter reflects the widening beam due to total reflection either at the glass surface or due to a radically decreasing refraction index (Fig. 2.18).

If near the source a beam is focused on the face of the fiber, the beam is guided to the end of the fiber by zig zag way due to total reflection (Fig. 2.18). If a ray in the Gaussian beam hits the face of the fiber under a certain angle (smaller than the divergence) at a certain radius (smaller than the beam radius) it takes a different zig zag path through the fiber than the beam edge (see Fig. 2.18) and reaches thus the end face of the

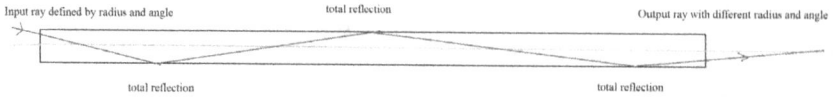

Fig. 2.18 Deviation from initially Gaussian beam after passing through the fiber. *Source*: Authors.

fiber at a different angle and radius and deviates thus from the initial position and direction in the Gaussian beam. Therefore the latter is not reproduced and a more complicated intensity distribution appears at the end of the fber.

Any deviation from straight linear geometry as for instance by bends leads to alterations of the transmission behavior as a reduction of transmitted beam power due to reflection in the fiber. Since a small part of the beam power is absorbed in the glass, heating and melting of the glass limits the capability of high power transmission to a multitude of kilowatts. Nevertheless the attenuation of the beam in the fiber is small enough to allow full transmission across several 10 m.

The fiber diameter must of course be larger than the focus size of the beam to avoid losses and is usually for technological applications in the order of magnitude of a few hundred micrometers dependent on transmitted power.

Practically the fiber is contained in a flexible hose. Due to the relative large diameter of the latter the fiber that is unfixed and loose can move freely in lateral direction in the hose and arrange itself to a smooth shape even if the hose experiences sharp bends. To be able to detect fiber break, a thin pair of metallic wires runs along the fiber and is short circuited at the end of the fiber. If fiber break occurs, the wires also are interrupted and indicate at the other end of the fiber a break thus allowing to stop laser feeding in order to avoid radiation leaving the fiber. On both ends of the light conducting cable two connectors are arranged that allow easy fixing of the cable to the beam source and to a processing head quite similiar to electric connections.

The Laser Center in Gmunden, that has been closed in the mean time for political reasons raping existing contracts although very succsessfull in terms of industrial developments operated a network driven by one

6 kW disk laser and distribution of the beam power by three fiber cables feeding a laser scanner, a 3D printing experimental setup and a robot, mainly for welding (Fig. 2.14).

2.6 Scanners for Fast Laser Processing

Due to the very low beam parameter product, solid state laser beams remain collimated along a considerable distance from the source. They can therefore be inclined by two rotatable mirrors in all directions and thus reach every point on a plane below the mirrors, usually the workpiece surface. The two mirrors, together called *scanner*, can electromechanically be turned one around a horizontal and one around a vertical axis (Fig. 2.19) and provide thus a high relative speed between beam and workpiece. The latter movement is much faster than one obtained with an

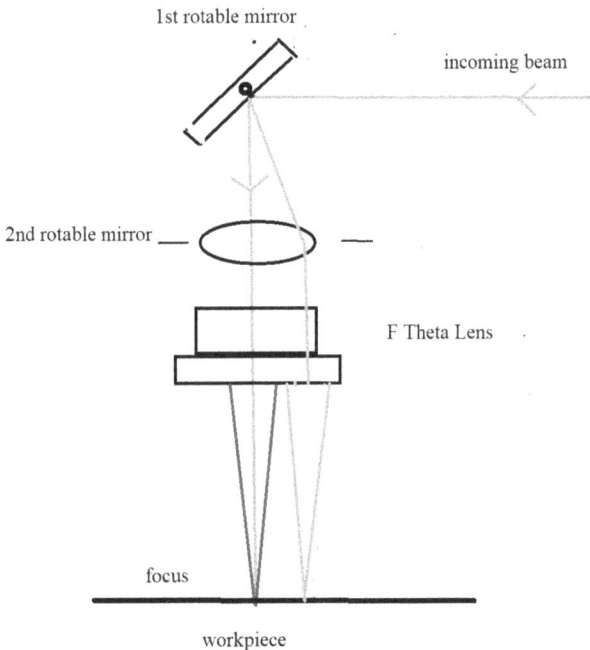

Fig. 2.19 Scanner with F-Theta lens.

Source: Authors.

xy table and can be used for material treatment where the unfocused beam is sufficient for heating as for forming, hardening or polishing.

If higher temperatures up to the evaporation point are necessary, especially for cutting and welding, the beam must be focused by a special optic, called *F-Theta lens*, that focuses the beam to a plane and keeps the focus size constant across the full working area after deflection. With the latter arrangement welding of multiple seams, each with a different starting point becomes faster due to quick repositioning of the beam.

F-Theta lenses consist of up to six individual lenses acting on the scanned beam subsequently, where the shape of the lenses has been determined by precise numerical calculations with the means of geometrical optics that assume light waves are simple rays that are refracted and reflected by the optical elements constituting the F-Theta lens.

The Laser Center in Gmunden operated successfully a laser scanner fed by a 6 kW disk laser via a fiber cable and an additional F-Theta lens mainly for hardening experiments, where a unique laser hardening process has been developed (Fig. 5.2). The latter uses *pyrometers* to measure the temperature in the focus of the laser beam and controls therewith the laser beam power in order to keep the temperature of the workpiece constant, even if distortions of the workpiece surface as holes or steps normally cause overheating due to obstructed heat conduction and thus surface degradation of the hardened workpiece.

2.7 Sensors for Process Performance

Sensors are necessary for the supervision of laser materials processing granting high workpiece quality and processing speed. Several phenomena are available to get the desired information about process performance as temperature, emission of light and sound and extrusion of molten material.

Of course temperature is the crucial parameter in laser material processing that always uses heat for softening of material in forming, material modifications as in hardening, melting in cutting and evaporation in welding and printing. The latter parameter can easily be measured with *pyrometers*, semiconductor devices that are dedicated to a specific temperature range between −30°C and 2,000°C. The latter sensors measure temperatures via the infrared radiation emitted by a hot material. They are focused to a small spot on the workpiece and collect the radiation preferably by a flexible light conductor finally feeding it to an IR sensitive semiconductor. The latter changes its electrical resistance due to the inner electrical photoeffect being a measure for the spot temperature. Special importance has temperature measurement for hardening where a relative narrow temperature range must be used.

Emission of visible light is crucial for all processes with lasers where not only the intensity of light emitted in the processed region but also the colors are of big importance. So red light is typical for hardening, yellow light is an indication for melting as in cutting with its lower temperature or white blue light is emitted in deep penetration welding that asks for evaporation (see Chapter 4).

The Vienna University of Technology Institute formerly headed by one of the authors developed already years ago a sensor based on lighting for cutting. Nowadays the Austrian company *Plasmo* offers sensors for welding based on the above considerations.

In rare cases laser cutting is accompanied by sound what can be explained with the oscillations of the volume of molten material always present in this process and can also be utilized to get indications about process behavior.

During laser cutting of metals a shower of molten and glowing droplets is ejected at the bottom of the workpiece thus performing the necessary material removal. Its observation, e.g., by an image processing system also can provide important information (see also Chapter 3).

References

1. Scott, Craig, Introduction to Optics and Optical Imaging, John Wiley and sons, 1998.
2. Eichler, H.-J., Eichler, J., Laser, Springer 2015.
3. *Fibre Bragg Gratings are periodic arrangements of thin lines or scratches on a fibre, that reflect light. All reflections from the lines are superimposed, whereby the distance between the lines, a fraction of the wavelength,*

determines the distinct wavelength that is reflected and not transmitted. Thus the bragg grating acts as a mirror.

4.   P. F., Moulton, *Ti-doped Sapphire: Tunable Solid-state Laser.* In: *Optics News, Vol. 8, Nr. 6.* 1982, S. 9–13.

5.   D. Schuöcker, G. Schiffner, Optimization of the diode-pumped YAG:Nd/+++/ fiber laser, Harvard Univ, 1978.

Chapter 3

Laser Processing with Material Removal

3.1 Overview

Drilling is performed with a focused laser beam that hits the workpiece without relative movement and heats thus the material continuously leading to a rising temperature soon reaching evaporation and consequently material ablation, then the latter forms the desired hole.

Cutting is performed by a focused laser beam that moves over the workpiece along the desired cut contour and heats and melts the workpiece. A sharply focused flow of gas either reactive as O_2 or inert as N_2 hits the molten material in the focus of the laser beam and leads to its ejection at the bottom of the workpiece. If reactive gas is used the mechanical action is enhanced by additional heating due to burning of the material.

Grooving is obtained by altering the direction of the gas jet from vertical as in cutting to nearly horizontal, thus moving melt ejection from the bottom of the workpiece to its surface. In consequence no through cutting is obtained but a groove is formed.

Laser ablation is performed by scanning a certain area on the surface of the workpiece line by line as mentioned above with sufficiently strong laser power to reach evaporation.

At lower laser power and weak evaporation thin layers of unwanted nature can be removed. With higher laser power shallow ponds can be created as for scribing, marking and engraving. Finally with strong laser power, deep excavations can be obtained as necessary for 3D shaping.

An even more efficient option concerns the abundance of evaporation and the use of melting alone due to eg enhanced speed of the laser beam that scans the surface of the workpiece line by line. In this case the melt must be removed by a nearly horizontal gas jet. This process is referred at as **laser planning**.

Also pulsed high peak power lasers as TEA or excimer lasers can carry out ablation without any movement of the workpiece by illuminating a mask and forming a shallow excavation with a single pulse which is an image of the mask. Subsequent pulses can be used to produce deep excavations for 3D shaping [2].

3.2 Laser Drilling

If a focused laser beam hits a workpiece, absorption heats the surface (Fig. 3.1). Since the bulk of the material is still cold, the temperature difference causes heat flow into the workpiece and thus a thin shell around the focus heats up and reaches soon the same temperature as in the focus on the surface. No heat conduction can take place any more. Nevertheless due to the permanent energy supply by absorption, the temperature at the surface must rise and then again a temperature difference is built-up between the surface and the shell mentioned above. Between the shell and the remaining bulk of the workpiece, that is still cold, also a temperature difference builds up causing heat flow. Thus shell by shell is heated whereas from shell to shell a temperature difference cares for heat conduction into infinity. Summarizing, continuous heating by absorbed laser beam energy leads to a rising temperature in the focus on the workpiece surface and also in the bulk of the material where the temperature decreases with rising distance from the surface, thus maintaining heat flow into the bulk of the workpiece. Eventually the temperature in the focus reaches a magnitude where evaporation sets on, what leads to very strong cooling since the deliberation of atoms from the solid material consumes energy. Finally the latter cooling compensates heating and the

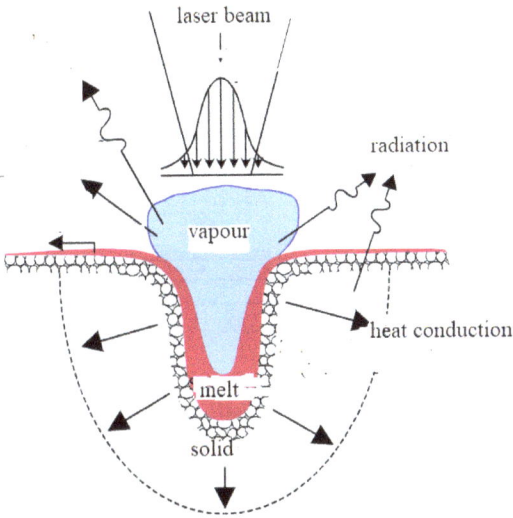

Fig. 3.1   Laser drilling (red: melt, blue: vapor).

Source: G. Chryssolouris, K. Salonitis Fundamentals of laser machining of composites, in Machining Technology for Composite Materials, 2012.

temperature in the focus remains constant although permanent heating by absorbed beam energy takes place.

The onset of evaporation causes ablation of solid material in the focus thus forming a hole that grows into the depth of the workpiece finally reaching its bottom.

Since normally evaporation is associated to melting, the walls and the bottom of the hole during drilling are covered by molten material. The latter is subject to the impact of the vapor pressure in the hole and is thus also driven out at the surface of the workpiece, enhancing the ablation process.

3.3 Laser Cutting

3.3.1 *Set up of a laser cutting machine*

The main constituents of a laser cutting machine are the laser, a focusing head with integrated gas supply and nozzle and a motion system that

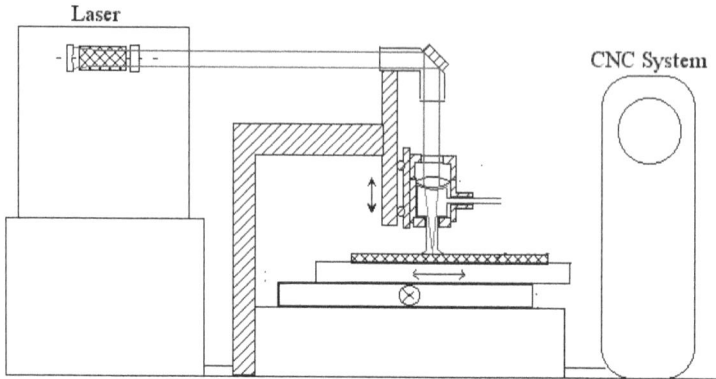

Fig. 3.2   Laser cutting system.

Source: D. Schuöcker, Spanlose Fertigung, Fig. 3.6.6 Oldenbourg, München 2004.

carries out the necessary relative movement between the laser beam and the workpiece (Fig. 3.2).

Lasers can be of CO_2 or solid state type with beam power up to many kW, pulsed or DC, preferably with a mode close to Gaussian since the latter yields a small focus and a narrow kerf. Solid state lasers allow a better absorption for metals combined with a small beam parameter product that provides a narrow kerf and large thickness of the workpiece and thus replace more and more the initially dominating CO_2 laser.

The beam generated by the laser is supplied via mirrors or fibers to the processing head that focuses it with lenses, in the case of CO_2 laser made from ZnSe, transmissive for far infrared, or for solid state lasers made of glass. The cutting head contains also the supply of gas for removal of melt, finally directed to the beam focus by a nozzle. Two kinds of gases are used: N_2 for mere mechanical action and O_2 for additional reaction with the workpiece that produces extra heat to enhance the cutting process. The purity of oxygen is very important for high cutting quality and speed. The cutting head ensures optimum perpendicular incidence. The latter avoids an elliptic deformation of the initially circular focus spot that would reduce beam intensity and thus cutting performance. The above system is completed by a loading and unloading facility for the raw material and the cut parts and finally a device that cares for collection and removal of the debris.

Fig. 3.3   Laser cutting and punching machine with a 3 kW CO_2 Laser and a workpiece held by clamps moving it in x and y direction across a resting table (developed by Voest Alpine, Linz/Austria 1983).

Source: D. Schuöcker, Spanlose Fertigung, Fig. 3.6.7 Oldenbourg, München 2004.

The workpiece is then moved with respect to the laser beam usually by a xy table that allows 2D cutting. If big parts-plane or 3D shaped-must be cut the workpiece is moved in one direction and the cutting head moves along a bridge perpendicular to the workpiece motion (*flying optics*). In the case of 3D cutting the focusing head can also move in vertical direction. To ensure perpendicular incidence at the workpiece as mentioned above, it can also be rotated around two perpendicular axles. Of course all five movements are CNC controlled (Fig. 3.4 and 3.5).

3.3.2 *Mechanism of laser cutting*

The laser beam hits the workpiece at the momentary end of the cut kerf, a nearly vertical area with striping incidence and is absorbed there strongly, according to the Fresnel mechanism mentioned in Chapter 1.

Thus the material is heated and a small molten droplet builds up. In the most important case of oxygen assisted cutting the melt consists of

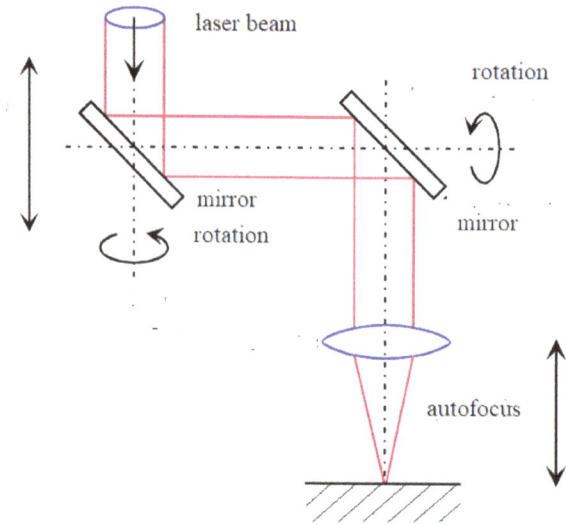

Fig. 3.4   Optical arrangement for 3D laser cutting.

Source: D. Schuöcker, Spanlose Fertigung, Fig. 3.6.8 Oldenbourg, München 2004.

Fig. 3.5   Gantry robot for 3D laser cutting (developed in Austria 1990).

Source: D. Schuöcker, Spanlose Fertigung, Fig. 3.6.9 Oldenbourg, München 2004.

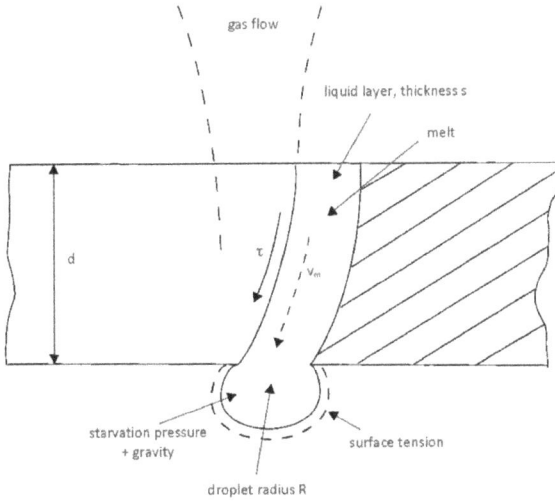

Fig. 3.6 Cross section of a workpiece partly cut, left view into the kerf, middle molten layer with closed loop mass flow, right part still uncut (LC Gmunden/Austria, 2012).

metal and slag. The latter is subject to friction with the gas jet formed by the nozzle in the cutting head and the molten mass in the droplet is accelerated. Nevertheless the melt cannot leave the latter since *surface tension* [2] prevents any mass leaving the droplet as long as the surface tension prevails the pressure in the droplet. Therefore a downward flow builds up in the droplet, that must reverse at its bottom thus forming a closed loop (Fig. 3.6).

The reversal of the melt flow at the bottom exerts a pressure, the *starvation* pressure [2] that reduces surface tension. Due to permanent acceleration of the melt flow by friction with the gas jet the melt speed and thus also the respective pressure rise and eventually the pressure equals the surface tension. At this point, the surface of the droplet breaks up and molten material flows out from the droplet thus removing material from the workpiece. After all the melt has been ejected a new period starts with building up of the liquid body and thus subsequent melt flow stops and goes. A numerical estimation yields for the duration of a period the order of magnitude of a second [3].

The highly dynamic and alternating nature of the molten droplet formed in laser cutting has two consequences.

Fig. 3.7 Steel, laser cut with typical striations (LC Gmunden/Austria, 2012).

D. Schuöcker, Spanlose Fertigung Fig. Oldenbourg, München 2004.
Source: Authors.

Fig. 3.8 Laser cutting of steel: melt ejection as spark shower (Spectra-Physics Germany)

Kindly submitted by Spectra Physics Germany (CEO Gukelberger, 1984) for public use.

First, it has an effect on the walls of the kerf, the cut edges, that are also molten in a shallow region that is in contact with the molten body, what leads to periodic striations typical for laser cutting and causing unwanted roughness of the cut edges (Fig. 3.7).

Second, molten material does not flow out from the workpiece continuously, but intermittent as a shower of glowing droplets, a spark shower, also typical for laser cutting (Fig. 3.8).

Independent from the melt removal mechanism as sketched above, the solid volume in the kerf must be molten to accomplish cutting. To ensure melting, the energy balance of the kerf volume scanned by the focused laser beam, consisting of **energy gain** by absorbed laser radiation and **energy used to heat** the volume of the kerf scanned per unit time and the heat lost by heat conduction must be in equilibrium.

Heat lost by the ejection of molten material is compensated by energy gain due to friction with the gas jet. Heat conduction can be neglected since the cut speed is usually high enough to prevent heat from penetrating into the depth of the workpiece. Thus the **heat gain** per second depends mainly on **absorbed laser power**.

The volume heated by the laser per unit time rises with the workpiece thickness, the cutting speed and the kerf width. The width of the kerf can be assumed to be approximately equal to the focal radius or a bit more because a much smaller width would mean that the walls of the kerf are not heated at all. On the other hand a much larger kerf width would mean that material is molten also behind the walls of the kerf what would lead to an increase of the kerf width. The **heat loss** for melting the kerf is now given by the specific heat that is needed to heat a unit volume by 1°C and the melting point and finally the product of focus radius, and most important, **workpiece thickness and cutting speed**.

Since heat gain and loss must be equal in equilibrium, the product of **thickness time's speed** depends mainly on **absorbed laser power**, a **relation that is crucial for all beam** processes. Of course the latter product depends also on the thermal properties of the material as well as on the focus radius, quantities that can be considered with a constant factor. Figure 3.9 shows measured and calculated values of speed and thickness that prove the above considerations.

3.3.3 *Practical cutting speed, thickness and quality for various metals*

Initially laser cutting has been carried out with carbon dioxide lasers with a beam power around 1,500 W and a focal size of 100 μm and more. For steel, a cutting speed around 10 m/min resulted, that could not be increased considerably by raising the beam power since carbon dioxide

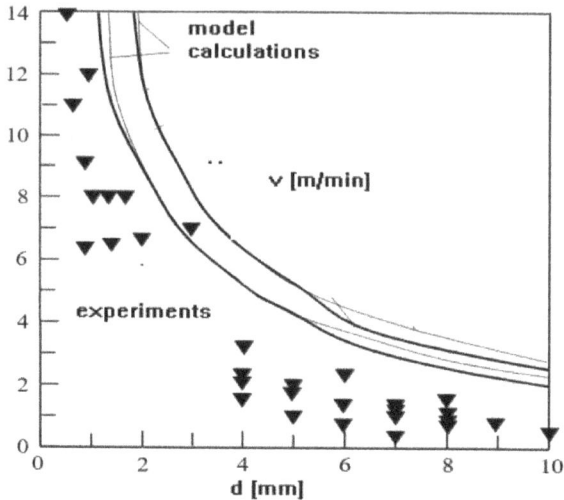

Fig. 3.9   Laser cutting of steel: speed vs. workpiece thickness calculated and measured[5] (Laser 1 kW, focus 200 μm).

Source: D. Schuöcker, Spanlose Fertigung Fig. 3.6.11, Oldenbourg, München 2004.

lasers showed a rising focus size at elevated beam power that did not lead to higher speed according to the above considerations. Nevertheless nowadays lasers are available with much higher beam power but still maintaining a small focus size as disk or fiber lasers, thus yielding speeds of 100 m/min and even more.

The workpiece thickness has a much narrower range limited by the gas jet, that cannot penetrate too much in the narrow kerf due to friction at the kerf walls, where practical limitations are around 15 mm for steel with the exception of special process options as the *Lasox* process (see below).

The kerf width is determined as mentioned above mainly by the focus radius and thus in the order of magnitude of 50–200 μm.

With lasers nearly every material as metals, plastics, glass and ceramics as well as fabric can be cut. Most important technical materials as steel, stainless steel and aluminum yield excellent results (see Figs. 3.10–3.12). Even copper, silver and gold can be cut satisfactorily with green laser light generated by frequency doubling of solid state laser radiation as mentioned in Chapter 2.

Fig. 3.10   Laser cut, construction steel 3 mm (TU Wien, 1985).
Source: D. Schuöcker, Spanlose Fertigung Fig. 3.6.12, Oldenbourg, München 2004.

Fig. 3.11   Laser cut of stainless steel 3 mm (TU Wien, 1985).
Source: D. Schuöcker, Spanlose Fertigung Fig. 3.6.13 Oldenbourg, München 2004.

Fig. 3.12   Laser cut of Aluminum 3 mm (TU Wien, 1983).
Source: D. Schuöcker, Spanlose Fertigung Fig. 3.6.14g Oldenbourg, München 2004.

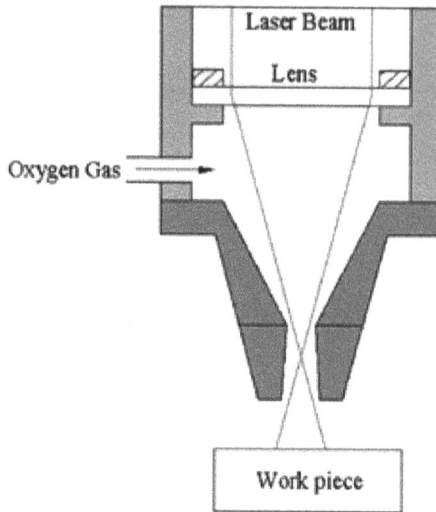

Fig. 3.13   Cutting head with laval nozzle.

Source: M. Sundar, A.K. Nath *et al.*, Int. J. Adv. Manuf. Techn. (2009)

3.3.4 *Cutting of thick sections*

With regular gas assisted laser cutting thick workpieces with a thicknessup to 100 mm and more in steel cannot be cut since the gas jet is not able to penetrate deeply into the narrow kerf although it is indispensable for cutting metals to remove the molten material. Only in the case of sublimating materials that evaporate without melting and with beams with very low divergence, that remain along a long distance well focused, thick materials can be cut. Nevertheless also thick metals can be cut with a special process that uses a laser only for igniting the burning of a metal in an intense oxygen jet and ejecting the resulting slag also by the latter jet (Fig. 3.13).

The laser beam hits the workpiece at the momentary end of the kerf with a relatively large focus spot and is absorbed there to reach the ignition temperature for steel (about 1,000°C).

Additionally an oxygen jet with a smaller cross section hits the workpiece in the focus of the laser beam. After the ignition of the burning

Fig. 3.14.   Lasox process 50 mm St, 1 kW Laser, v = 0,275 m/min.

Source: M. Sundar, A.K. Nath *et al.*, Int. J. Adv. Manuf. Techn. (2009)

process, additional heat is generated by the burning process and the melt-ing point (<>1500° for steel) is easily reached and metal and slag are molten. The resulting melt is ejected due to friction with the gas jet at the bottom of the workpiece.

In this process the kerf width is determined by the oxygen jet not by the laser beam as in normal cutting and is also much larger and therefore the gas flow can easily reach the depth of the workpiece.

To get a wide focus of the laser beam that ensures ignition of burn-ing across the cross section of the gas jet and also to get a high gas speed to deliver oxygen at a large rate, practically in the *supersonic* region, a *Laval nozzle* is used in the cutting head (Fig. 3.13). The latter nozzle has an inner diameter that decreases to a narrow opening and then widens again to obtain supersonic flow speed. The focus of the laser beam is located at the waist of the nozzle, thus forming the necessary large spot at the workpiece.

Fig. 3.15   Laser engraving of a stamp for the production of medals (TU Wien, 1990). *Source*: D. Schuöcker, Spanlose Fertigung Fig. 3.6.4 Oldenbourg, München 2004.

3.4 Grooving

If for given absorbed laser power and focus width the cutting speed is rather high the full thickness of the workpiece cannot be reached and the kerf can only be molten to a certain depth within the workpiece. In this case the ejection of molten material cannot be carried out by a vertical gas jet but must be performed by a nearly horizontal one that blows the melt in the forward direction.

Width and depth of the resulting groove can be adjusted by changing focus size and speed. The latter process resembles planning and is thus called laser planning.

The latter process can be used for scribing and marking as well as facilitating bending by quasi reducing the thickness of the workpiece along the intended bending edge and thus reducing the bending force and enhance the maximum thickness to be bent. By arranging a multitude of grooves shoulder to shoulder also areas can be processed, e.g., to generate a 3D structure with variable depth.

3.5 Laser Ablation

This 2D process is in principle performed by scanning the area to be ablated with a laser zig zag line by line. Material removal is accomplished either by **melting** the material to a desired depth and ejection of the melt by a nearly horizontal gas jet as in laser grooving as above or by **evaporation**. In the first option the strong friction between gas flow and work piece causes turbulences that cause a chaotic flow of molten droplets that are not all exhausted but resettle on the workpiece. Therefore the resulting surface is spoiled with resolidified melt. Also incomplete melt removal leads to a low quality. Nevertheless the processing speed is high and the depth of ablation can be easily adjusted in a wide range and therefore 3D structures can be created.

With better focusing, higher beam power or lower speed the second option of evaporation can be realized by reaching temperatures as high as the vaporization point, preferably with pulsed radiation.

In this case the material in and around the focus is molten and even starts to evaporate leading to material removal as vapor. The latter exerts a recoil force on the molten body in the beam focus and prevent the latter from chaotic ejection as in the first option, even more since no ejecting gas flow is used. Therefore clean surfaces are obtained. If the beam is pulsed the heat has not much time to flow away from the molten and evaporating region and thus the processed regions have sharp borders corresponding to the scanned area. The latter option is used by industrial engraving machines as shown in Fig. 3.16.

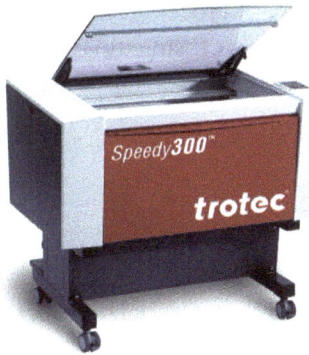

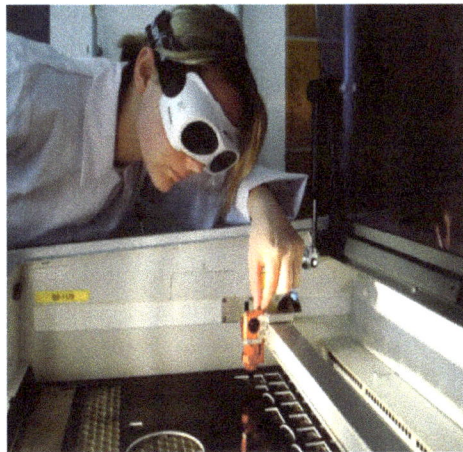

Fig. 3.16. Laserplotter developed by TU Wien and Trotec, Wels/Austria Right: gantry system and cutting head Kindly permitted by Trotec Comp, Wels/Austria.

The latter machine is connected to a PC that allows to create a foto, drawing or text and to hand it over directly to the plotter for engraving on nearly any material desired.

Besides generation of shallow 3D structures also surface cleaning, e.g., from rust is an important application, where thin layers are removed.

Several thousand machines of the above kind have been sold all over the world, thus justifying Trotec/Wels Austria to be world leader in engraving machines as described above.

The first option of melting and melt ejection by a gas jet is the reverse process to 3D printing where molten material is deposited line by line on a workpiece thus building up structures (see Chapter 4).

A third option concerns the use of masks with an opening that corresponds to the shape of the area to be ablated. A strong laser pulse with a cross section that covers the whole opening can now be used to ablate material across a desired region given by the said opening. If short laser pulses are used heat cannot flow away and sharp borders of the ablated regions are obtained. For the latter purpose preferably Excimer lasers are used (see Chapter 2).

Comparing all three options, ablation by mere evaporation with a moving laser beam is fast and provides a quality superior to that obtained with melt ejection by a gas jet. Other than the mask process it also allows it to process large areas, although with larger processing time and is cheaper since mask production is not necessary.

References

1.   Poprawe, R., Lasertechnik für die Fertigung, Springer 2005.
2.   Wikipedia-the free encyclopedia, Fluid mechanics, 2020.
3.   Dynamic phenomena in laser cutting and process performance D Schuöcker, J Aichinger, R Majer — Physics Procedia, 2012 — Elsevier.

Chapter 4
Processes with Material Addition

4.1 Overview

Laser welding is performed by moving the parts to be joined to each other until they nearly touch, leaving a narrow gap. The rim and the walls of the latter are molten with a laser beam, until they form a common melt pool, that becomes the weld seam after resolidification.

3D printing is used to build up parts by subsequent creating thin layers that correspond to cross-sections of the work piece under construction by welding down metal powders or wires with a laser.

4.2 Laser Welding

4.2.1 *Layout of laser welding systems*

A laser welding facility (Fig. 4.1) consists similar to cutting of a laser source, a processing head and a movement unit that allows the laser beam to move along the gap between the parts to be joined. Different from cutting is the welding head, that does not use transmissive focusing optics as lenses, that can not withstand the higher beam power necessary for welding, but concave focusing mirrors made from metals, which are easily cooled by water flows. Moreover, often robots are used for moving the welding head with respect to the workpieces, especially if they are large, a solution that is not really feasible for cutting due to vibrations

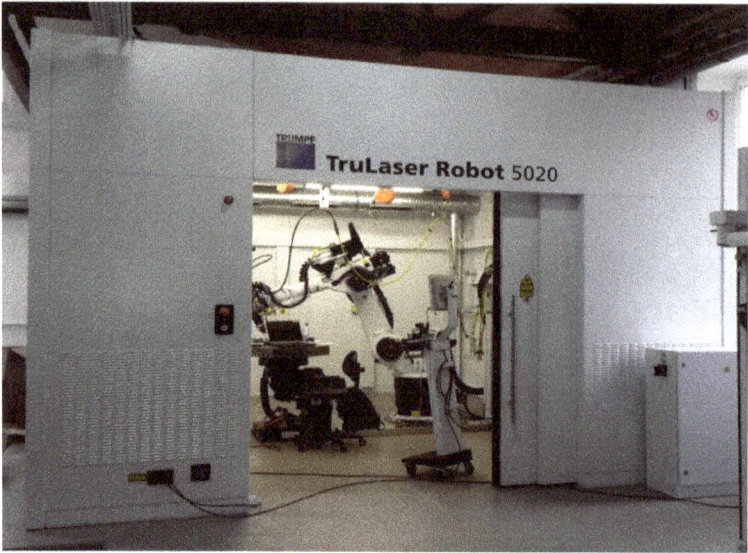

Fig. 4.1   Welding with articulated robot and laser (LC Gmunden/Austria 2015).
Source: Authors and coworkers.

superimposed to the movement. Also specific fixing tools must be used to prevent deformations of the parts to be joined due to stress caused by high temperatures. Finally, in some cases as welding of aluminum filler wires must be supplied to avoid the formation of pores (Fig. 4.2).

4.2.2 *Mechanism of laser welding options*

4.2.2.1 *Heat conduction welding*

The energy supplied by the laser beam heats the surface of the two work-pieces and flows in all directions of the workpieces thus leading to a decreasing temperature with rising distance from the surface. The isother-mals are half globes around the center or the laser spot. The one associated to the melting point limits the molten region. The extension of the latter determines not only the width of the weld seam but also its penetration depth, thus being the maximum thickness that can be joined, practically in

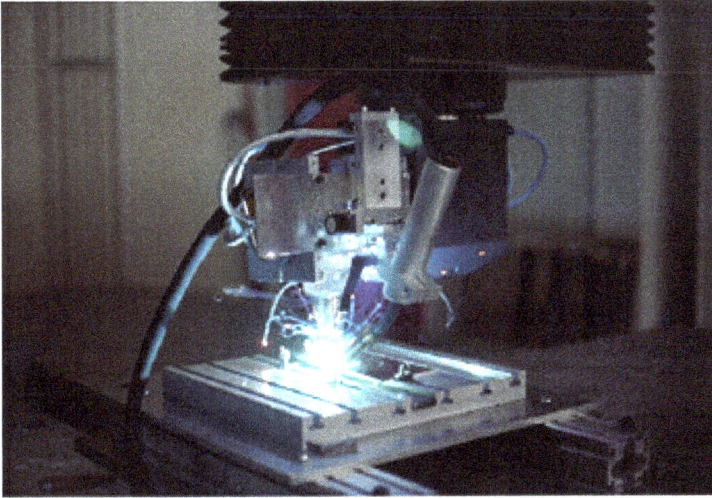

Fig. 4.2 Laser welding of aluminum with filler wire (TU Wien, 2000).
Source: Authors and coworkers.

the order of magnitude of 1 mm for steel and depends of course on the absorbed beam power. The weld seam cross-section is wide at the surface and has a half circular shape becoming smaller or nearly zero at the bottom of the workpiece. The strength of the seam exceeds that of the virgin metal due to the remelting process, but for the same reason an enhanced hardness appears with the threat of cracks during cooling down or forming (see Chapter 12). A heat affected zone forms in the vicinity of the seam also showing modified microstructure with increased hardness (*heat affected zone HAZ*) (Fig. 4.3). During the welding process the unjoined gap in front of the welding laser beam tends to widen due to the thermal expansion of the workpiece edges. Since the bulk of the workpieces remains cool, the edges incline towards the latter to keep the workpiece volume constant, thus widening the gap and obstructing further welding. To overcome that problem, sophisticated fixtures must be used to avoid the latter widening. Also a chain of pointlike weldments prior to seam welding can prevent the widening of he gap and allow subsequent welding.

Fig. 4.3   Paper filter metal cover heat conduction welded Kindly submitted by Spectra Physics Germany (CEO Gukelberger, 1984) for public use.

Heat conduction welding is quite useful in welding of car bodies and suppliances and is well suited for use of diode direct lasers in the kW range since the larger focus and divergence are tolerated.

4.2.2.2 *Deep penetration welding*

If the workpiece thickness becomes larger, heat conduction welding can only be performed with big seam width to obtain a deep molten body and thus a modified process is necessary.

The latter uses not only melting but also evaporation and asks thus for much stronger lasers than for heat conduction welding.

Evaporation forms a hole with a diameter similar to the focus size (Fig. 4.4). The latter is filled with vapor and vapor also leaves it at the surface of the workpiece. If the laser is strong enough to ionize the vapor and convert it to a plasma that shows perfect absorption for laser radiation in the IR region as of CO_2 and Nd:YAG lasers and couples it to 100% to the workpiece and makes the process very efficient. The onset of this process mode is associated with a bright light blue light emission (Fig. 4.2). The vapor channel is surrounded by a molten region that extends laterally in both workpieces. In forward direction the melt is short since there is a high temperature gradient because the melt pool reaches cool material that must be heated. On the back side the melt is more extended since there is a low temperature gradient because hot regions are left. Due to the motion

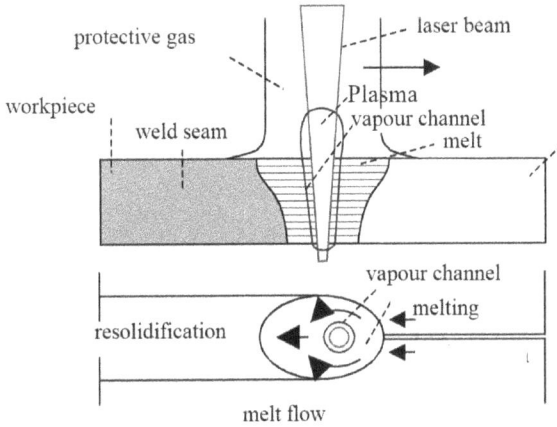

Fig. 4.4 Mechanism of deep penetration welding.

Source: Authors and coworkers.

of the melt pool in welding direction at its front side solid material is molten and flows around the keyhole. At the back side melt accumulates and cools and resolidifies thus joining the two workpieces by a weld seam.

Speed and maximum thickness of deep penetration welding are comparable to cutting, since in both processes the material must be molten throughout the depth of the workpiece and therefore the role of the product of both quantities applies also. Nevertheless often considerably high beam power as provided by CO_2 or solid state lasers of many kilowatts is necessary.

The heat input to the workpiece is smallest of all welding processes with the exception of E beam welding and therefore deformations of the workpiece during welding are less likely with the exception of gap widening as mentioned above [2, 3].

4.2.2.3 *Laser hybrid welding*

In laser hybrid welding, the heat sources for melting the material are a high power laser beam on the one hand and an electrical arc on the other hand. By the combination of two energy sources, the welding depth like the welding speed will be increased considerably. In detail, the

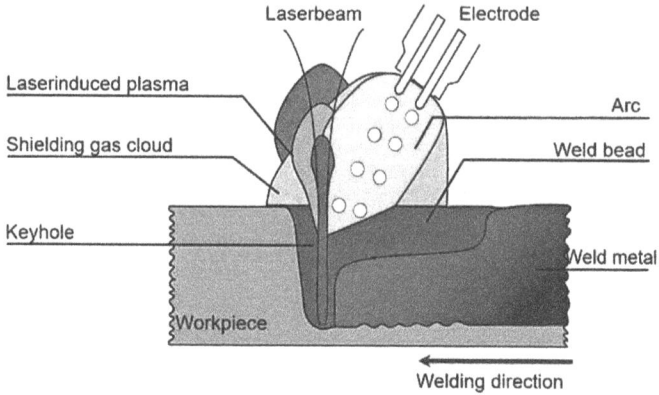

Fig. 4.5   Mechanism of laser hybrid welding. With kind permission of Fronius, Wels/ Austria.

well-focused laser beam creates a narrow weld seam, while the much thicker arc ensures good gap bridging and therefore the requirements for the seam preparation are less stringent.

Apart from these advantages the need for shielding gas will also be considerably reduced.

The latter mechanism of hybrid welding is illustrated in Fig. 4.5.

Welding with a robot guidance of the welding head, which on the one hand performs the feed of the wire, which is used as electrode for the arc and on the other hand carries the focusing optics for the laser beam as well as the protective gas supply, is especially suited for thick aluminum parts.

4.2.2.4 *Laser remote welding*

As already treated in Chapter 2 a scanner with two rotatable mirrors and a large, special lens of F-Theta type can be used to move a laser beam with high speed over a large work area, especially if a robot guidance of the scanner is used. Requirement for using this technology is a high quality of the laser beam, such as that from disk or fiber lasers.

The particular advantage of this procedure is that the laser beam with almost no inertia can be moved over the work piece at high speed and thus for positioning at the welding point very short times are sufficient, what

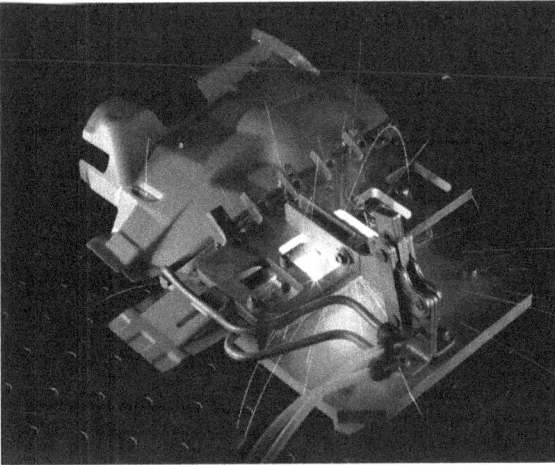

Fig. 4.6   Remote welding of a car part.

Source: D. Schuöcker, Neueste Entwicklungen der Lasertechnik, METALL Monatsschrift des Wirtschaftsverlages Wien, 2007.

makes remote welding well suited for interrupted welds or for work pieces with multiple welds and then the utilization of the laser amounts to almost 100% (Fig. 4.6).

4.2.3 *Welding defects*

Pores and cracks may be considered as welding defects [1]. Although cracks can develop at almost any temperature they are classified as hot cracks and cold cracks.

4.2.3.1 *Pores*

If the seam material contains atomic gases, as for instance hydrogen, stemming from spoiling as by oil or fat, and cooling down of the weld seam takes place rather quickly, the gas atoms have no time to leave the weld seam and are practically trapped in the latter, and assemble in pores. Usually weld seams contain many pores with extensions of no more than 0.2 mm, but sometimes these pores are larger. Industrial standards state maximum size and number per unit area of the seam cross-section for laser welding of specific materials, as for instance aluminum, since the formation of pores is important concerning the strength of the welding. Personal experience with a project of laser welding of aluminum has shown that without special precautions the number and size of pores in the weld seam exceeded the allowed limit. Experiments at TU Vienna have been carried out to reduce the number of pores. Two measures were successful, namely the use of additional aluminum wire (see Fig. 4.2), because the latter seemed to fill up pores, and supporting the bottom of the weld seam in order to avoid molten aluminum flowing away. With these measures the standards mentioned above could be met.

4.2.3.2 *Hot cracks*

Resolidification of the molten material between the two workpieces that will later on form the weld seam will start from both sides of the weld seam and will proceed towards the center of the seam. During this resolidification the material cools down and shrinks on both sides of the seam, leaving a gap in the center of the seam that is filled up by liquid material flowing in from the molten body around the weld source. If for instance this supply with liquid material is obstructed due to a narrow and deep seam geometry, then the supply of this liquid material is incomplete, and a crack remains in the middle of the seam extending in parallel to the latter. These cracks can not only appear in carbon steels, but also in austenitic ones, aluminum and other important metals.

From this explanation of the formation of hot cracks, the circumstances leading and furthering the latter — such as high welding speed, large differences between solidus and liquidus temperature (see Chapter 9), and also a narrow geometry of the weld seam — can easily be understood. It is also known for experiments that materials with high sulfur and phosphate lead to hot cracks, where the presence of manganese improves the situation.

Measures to avoid hot cracks can easily be derived from the above facts, including (besides improvement of weld geometry) reduction of welding speed and heating of the welded workpiece, since this facilitates the flow of liquid material in the gap formed during resolidification, as explained above.

4.2.3.3 *Cold cracks*

Cold cracks are formed mainly at room temperature in the weld seam and in the heat-affected zone (HAZ). If, on the one hand, cooling down after welding takes place quite fast, also supported by a high content of carbon, a Martensite structure (see Chapter 9) is achieved. On the other hand, hydrogen being present in the welding parts is a prerequisite for cold cracks.

The mechanism of the formation of cold cracks is now the following. Hydrogen diffuses during the cooling phase through the material to assemble in bubbles, and the latter of course enhances the stress stemming from cooling down and shrinking the material in the weld seam and the HAZ. If now, due to high carbon content and fast welding associated with quick cooling, a Martensite structure is formed, the enhanced stress can lead to ruptures and cracks that can extend in all directions in the seam and the HAZ. These cracks lead always to a complete failure of the welding.

The latter cold cracks can be avoided by a low carbon content with a maximum of 0.2%. Heating prior to welding also helps, because then cooling down is slower after welding. Finally, also annealing at 250°C helps since it reduces stress induced by the welding process. These measures can prevent the formation of Martensite structures. So, it is not difficult to avoid the formation of cold cracks with careful planning of the welding process.

4.2.4 *Weldability*

4.2.4.1 *Steel*

In laser welding of steel, the main problems are cold cracks caused by embrittlement of the weld seam and the HAZ due to the formation of

Martensite structures and also gas atoms present in the material. It is therefore important that the carbon content that is responsible for embrittlement remains below 0.3% and that no gases such as oxygen or hydrogen are present. Also, sulfur and phosphate make the steel brittle and should be avoided. Other impurities such as chromium, manganese, vanadium, molybdenum, copper, nickel, and others give the steel improved performance and exert only weak effects on the welding process. The latter can be described by the carbon equivalent:

$$CE = C + Mn/6 + (Cr + Mo + V)/5 + (Ni + Cu)/15 \, (\%)$$

If the latter equivalent is lower than 40–50%, good weld quality without cracks can be expected.

4.2.4.2 *Stainless steels and cast iron*

Stainless steels with chromium content up to 20% and austenitic crystallization have of course a tendency to form martensite that makes cracks likely. On the other hand, intercrystalline corrosion plays not a big role in laser welding due to a favorable temperature range [2]. Anyways, preheating and also annealing after welding help to avoid martensite formation and thus can avoid cracks.

Cast iron can also be welded with lasers, but due to its strong brittleness preheating and annealing are obligatory.

4.2.4.3 *Aluminum and its alloys*

As already mentioned in the subchapter on pores, the latter are an important problem. Not only hydrogen, but also evaporated magnesium and zinc, the most important alloy elements of aluminum, can cause unavoidable pores.

Precipitation hardened aluminum can be softened during laser welding on the HAZ. Thus hardening should be performed after welding, not before. So, good results an be achieved with laser welding if some helpful measures are used.

4.2.4.4 *Welding of surface treated steels*

Zinc coatings leads to small pores in the weldment due to its low evaporation temperature that generates vapor bubbles penetrating the molten material. These pores do not essentially degrade weldment quality.

Nitrided steel yields only poor laser welding results, caused by diffusion of nitrogen in the weldment that leads to increased strength and decreased toughness, and creates holes in the weldment surface.

Nickel-coated steel cannot be welded with lasers, according to experimental investigations.

4.3 3D Printing

4.3.1 *Overview*

3D printing, that means the production of the work piece layer by layer mainly from powderized or wire shaped metal, is somewhat superior to conventional machining since the workpiece shape is not achieved by removing material but by building up the workpiece, saving material and related energy. By the way no tools are necessary and thus nothing suffers from any wear, in summary the process is very economic.

Since the building process takes place by adding small amounts of material with properties determined by software, the building process can be numerically controlled with high precision. Thus the machine can be directly be fed by a CAD program or by data sent via email. The printing process does no harm to the environment since no vapor, dust or powders are emitted and also noise is omitted. It should also be mentioned that 3D printing can be used for nearly any material besides all metals also plastics (thermoplasts), ceramics, glass and many others.

Shortcomings are that the building process is rather slow and the surface quality of the produced parts is limited by the grain size of the powder used as base material.

Various options can be chosen, where the most prominent uses a pre-placed powder bed that is selectively molten down with a laser beam and can be applied to any material. Instead of a powder layer also powder can be blown into the melt pool on the workpiece formed by the laser beam. Finally wires or filaments can be supplied to the melt pool.

The latter options will be treated in the bulk of the chapter with special attendance to metals and plastics.

4.3.2 *Selective laser melting (SLM)*

4.3.2.1 *Layout of a 3D printer*

The 3D printer (Fig. 4.7) primarily consists of an open container, in which the construction platform is located. The latter can be moved CNC controled up and down. The container is completely filled with metal powder, whereby the construction platform is placed in such a way that it is covered with thin layer of powder. A focused laser beam now scans the desired cross-sectional layer of the workpiece under construction CNC controlled line by line and melts the powder in the process, so that it is welded with the foregoing layer. After completing this laser

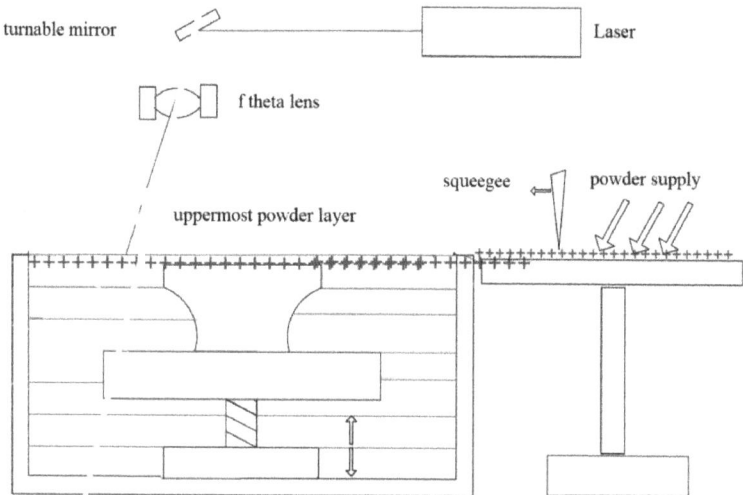

Fig. 4.7   Layout and function of a SLM machine.

Source: Authors.

Fig. 4.8   Powder bed preparation SLM process (Shutterstock purchase).

melting process, the construction platform with the workpiece under construction moves down by a layer thickness and then the next thin powder layer is applied with a squeegee as shown in Fig. 4.8.

The necessary relative movement between laser beam and workpiece is performed by a scanner as already treated in Chapter 2. The latter is basically a CNC-controlled mirror turned around two axes, which deflects the laser beam in such a way that every point on the construction platform can be reached very quickly. A prerequisite for this beam guidance is the use of a laser with a low divergence. A f theta lens is downstream of the scanner and focuses the laser beam on the construction platform and ensures that the focus remains on the construction platform at every scanner angle. In order to avoid oxidation of the metal powder, the printing process takes place in a vacuum-tight chamber, which is filled with a protective gas, such as argon (Figs. 4.9–4.10).

4.3.2.2 *Building rate of the SLM process*

Similar to laser cutting, an energy balance can also be established for the melting down of metal powder. This includes first the net energy gain, as absorbed laser power minus the losses by heat conducted to the bulk of the workpiece.

Losses due to heat conduction must be proportional to the temperature of the layer, the melting point, and also the thermal conductivity and the

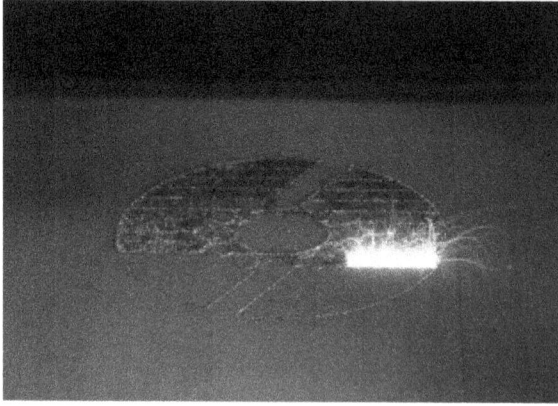

Fig. 4.9   SLM production of a metal grid (Trumpf, Ditzingen/FRG).

Fig. 4.10   Complex steel part manufactured by SLM Wth kind permission of Trumpf, Ditzingen/BRD.

focus radius, since with a larger melting zone the heat dissipation is greater. Finally the traveling speed together with a time constant describing the speed of heating in the previously built layers must be taken into account (net heat gain see Fig. 4.11).

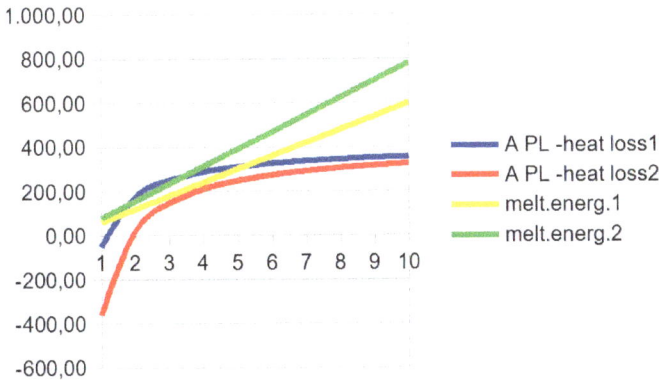

Fig. 4.11 Net heat gain and melting heat (W) vs. scanning speed (m/s) for two different focus radii.

Source: Authors.

The energy necessary to heat the powder under the track of the scanning laser up to the melting point is given by the width of the track, the thickness of the powder layer, the scanning speed and the specific heat (the amount of heat required for heating 1 m^3 by 1°C) and finally the melting point (melting heat), see Fig. 4.11.

As an example Fig. 4.11 shows the net heat gain and melt power dependent on speed in m/s for steel with an absorbed laser power 0.4 kW and two focus radii $r_{Focus1} = 1E\text{-}4m$ and $r_{Fokus2} = 1.3E\text{-}4m$.

The net power gain is the difference between absorbed laser power and heat losses into the bulk of the workpiece and can be used for melting the thin powder layer on the surface of the workpiece. The latter quantity as shown in Fig. 4.11 depends strongly on the scanning speed. At speed zero (no laser movement) all the absorbed laser Power flows into the depth of the workpiece (see drilling) and thus the net gain is zero. With rising speed heat has less time to penetrate into the workpiece and thus the net gain rises. For very high speed no heat can penetrate into the material and the net gain becomes equal to the absorbed laser power. The latter quantity also decreases with growing focus since then more heat flows out of the focus in the bulk of the workpiece. Of course the net energy gain rises with rising laser power.

The second quantity, the energy necessary for melting of the scanned powder rises with rising speed as shown in Fig. 4.11. Since then more powder must be molten per unit time and also must rise with rising focus size since then the molten trace is wider.

Figure 4.11 shows that these curves have two intersections for a certain focus radius and no intersections for a larger focus. Thus the actual focus radius is determined by the situation that both curves touch each other, what also yields the resulting processing speed.

In the example used here the resulting focus radius is 1.15E-4 m and $v = 2.8$ m/s (Fig. 4.12). The building rate is now given by the cross-section of the molten and resolidified track, the product of the layer thickness and its width times the distance scanned by the laser per unit volume, the speed. If the thickness of the layer is assumed to be 0.1 mm and the layer width is approximately equal to the focus radius as determined above together with the speed also determined above a building rate $dV/dt = 117$ cm^3/h results, a value with an order of magnitude found in experiments. It points out that the latter rises with rising focus size, a favorable performance but associated to strongly reduced quality due to a skipping of fine details of the workpiece shape.

If one now neglects the heat conduction with the argument that the momentary surface of the workpiece in production must be only slightly below the melting temperature, one obtains a relationship for the addition rate, in which the latter is proportional to the absorbed laser power. Since the latter, i.e., the produced volume per unit time, is composed from travel

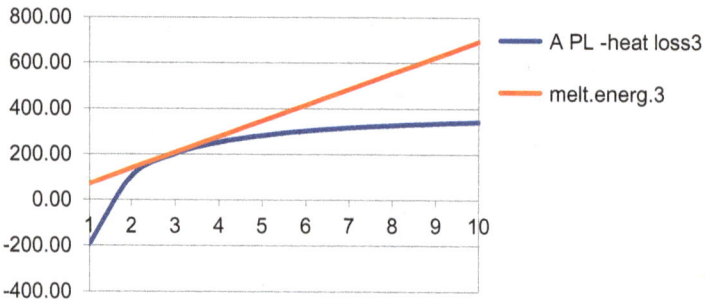

Fig. 4.12   Net heat gain nd melting heat vs. speed for the resulting focus.
Source: Authors.

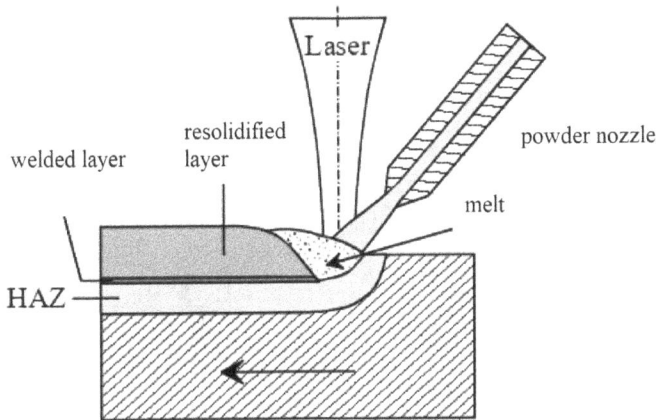

Fig. 4.13 Mechanism of the LMD process.

Source: Process control and further improvements for the blown powder process M. Resch, I. Smid, and D. Schuöcker: ICALEO 2001, 722 (2001).

speed, layer thickness and layer width, the relationship already found for laser cutting and welding and stating that the product of travel speed and layer thickness only depends on the absorbed laser power and material constants, such as melting temperature and specific heat also applies here. In very rough approximation, this relationship seems to apply to all kinds of laser material processing.

4.3.3 *Laser metal deposition (LMD)*

4.3.3.1 *Layout of a LMD machine*

In the actual process, one or more nozzles direct a beam of powderized metal into the focus of a laser beam, melting it in flight and settling it in liquid form on the workpiece surface (Fig. 4.14). Due to the movement of powder nozzle and laser beam over the workpiece surface, the material cools down and resolidifies after the laser beam has moved on and is welded to the momentary surface of the workpiece under construction.

This operation only takes place in the desired workpiece cross-section. The latter method, LMD (Laser Metal deposition), has compared to the powder bed process discussed the advantage of almost any size of the

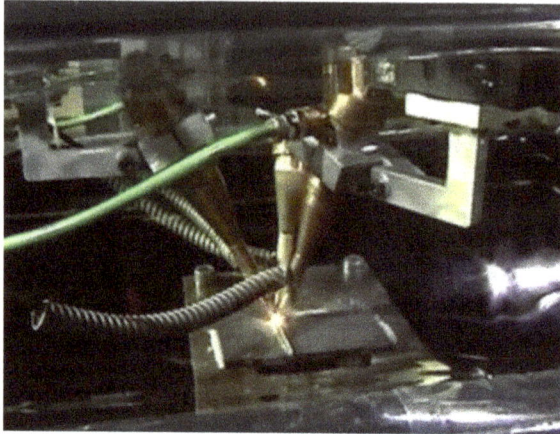

Fig. 4.14 LMD machine with three powder nozzles and central laser beam.
Source: Authors.

workpiece. Nevertheless the necessary enclosure that protects the opera-
tors against the threat of lung diseases due to inhaled powder makes is
more difficult. Initial attempts to develop this process were successfully
performed at the Institute for High Power Laser Technology at the Vienna
University of Technology about 20 years ago together with the Seibersdorf
Research [4].

In these experiments the nozzle diameter was 1.5 mm, inclined by 30°
towards the workpiece. Steel powder Böhler W300 was used with grain
sizes 40 to 150 μm. Laser power was 1 kW. Powder feeding rate below 5
g/min yielded very smooth surface, powder delivery 5 and 20 g/min a
slightly increased roughness and feeding rates above 20 g/min high building rates and very high surface roughness. As in all layer based printing
processes the tensile strength differ in the direction in parallel to the layers
and in a perpendicular direction by 30%. The calculation of the building
rate is quite similar as shown above for the SLS process with the only difference that the product of layer depth and width in the SLS analysis must
be replaced by the cross-section of the nozzle, what does not change the
magnitude of the building rate of 100 cm^3/h. In order to achieve direction-
independence, three powder nozzles (Fig. 4.14) have been used. As a more
elegant solution, a coaxial nozzle with a central laser beam and a surrounding annular powder flow was later developed (Fig. 4.15).

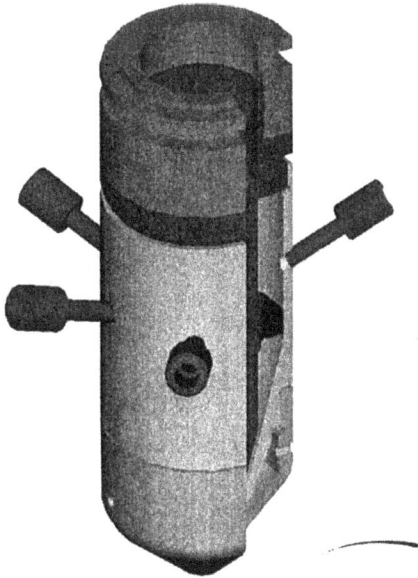

Fig. 4.15 LMD nozzle with central beam, coaxial powder and protective gas. With kind permission of Austrian Institute of Technology.

4.3.4 *Laser wire deposition (LWD)*

This process uses thin wires instead of metal powder that is fed to the printing nozzle by a storage spool and moved by two wheels with opposite rotation directions due to friction.

Advantages are the absence of powder that could be harmful, a hundred percent use of the base material and a higher building rate due to the higher density of wires compared to a powder flow and easy handling of the supply with the base material since no transport gas is necessary and the wires are simply moved by friction with two wheels rotating in opposite directions. Attempts to develop this process have been made at the Laser Center Gmunden/Austria with a disk laser with maximum power of 6 kW fed to the experimental setup by a glass fiber. The focused laser beam heated and melted a steel wire with submillimeter diameter and thus molten droplets settled at the workpiece (Figs. 4.16 and 4.17).

An industrial wire supply, laser melting and deposition head has been developed by Fraunhofer Institut für Lasertechnik Dresden with a central wire feed and laterally symmetrical beam incidence (Fig. 4.18). The latter

Fig. 4.16   LWD process (LC Gmunden 2015).

Source: Authors.

Fig. 4.17   Wall built by wire deposition (LC Gmunden 2015).

Source: Authors.

device reaches a building rate of 3 kg/h and a high spatial resolution due to a very low wire diameter less than half a millimeter. As shown in the figure, the wire deposition head is usually carried by a robot, thus enabling the processing of large parts.

A competitive solution with a central laser beam and three wires fed laterally towards the laser beam has been proposed and patented by the

Fig. 4.18   Wire deposition head developed in Dresden.
With kind permission of Fraunhofer Institut für Werkstoff und Strahltechnik (IWS) Dresden.

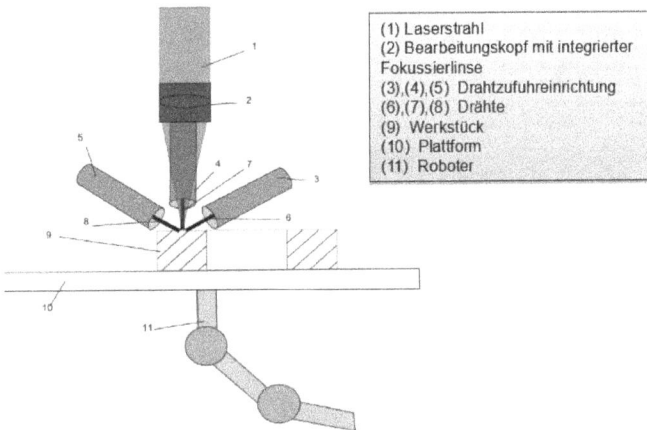

(1) Laserstrahl
(2) Bearbeitungskopf mit integrierter Fokussierlinse
(3),(4),(5) Drahtzufuhreinrichtung
(6),(7),(8) Drähte
(9) Werkstück
(10) Plattform
(11) Roboter

Fig. 4.19   Scheme of a wire deposition system with variable building speed and spatial resolution.
Source: Authors.

Laser center in Gmunden (2015, see Fig. 4.19) [5]. The latter arrangement has the advantage that all three wires could be controlled independently thus allowing to use only one of them for a small melt pool that grants high spatial resolution but reduced building rate or using all three wires for high building rate with lower spatial resolution. The latter concept allows to optimize the building of parts with regions without many details and highly structured regions. That idea seems quite promising, but the regional government declined promised financing for pure political reasons.

4.3.5 *Printing of polymers*

The difference between printing of metals and of plastics is essentially the much lower melting point of the latter (10% for acrylic glass) and the much larger thermal expansion (10 times). Due to the first item electrical heating is sufficient. Nevertheless for very high building rates as wanted laser could become necessary and therefore plastics are treated here also.

Concerning the second item, deformations from the intended shape are likely due to the strong thermal expansion that leads to delaminations during the building process. Both items are treated below.

A printer for such materials (Fig. 4.20) consists essentially of the printing platform on which the workpiece is built and the extruder which melts the initially filamentary or granular material and supplies it through

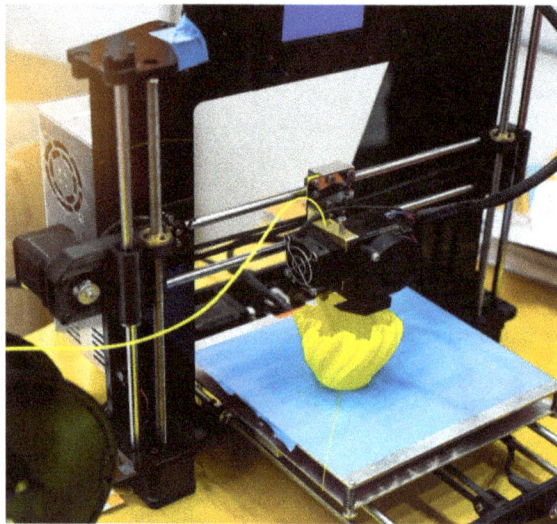

Fig. 4.20   3D printer for thermoplastic filament.

Source: (Shutterstock purchase)

a nozzle to the workpiece and finally a movement device that moves the extruder over the workpiece. In addition, of course, there is a CNC control and software, whereby data supplied by a CAD program about the geometry of the workpiece are available as a ".stl" file. The latter is then converted in a second program, the slicer, in a printable version with some extras as means to reduce weight of solid regions (honeycomb structure) and then a program for controlling the printer, which essentially contains the G commands of CNC.

Extruders similar to screw extruders (screw type) used in injection molding are already available on the market for the processing of cheaper granules that can simply be produced by milling plastic pieces (Fig. 4.21). Another possibility to realize an extruder with high material throughput is that the raw material in a heated cylinder is compressed and relaxed by an

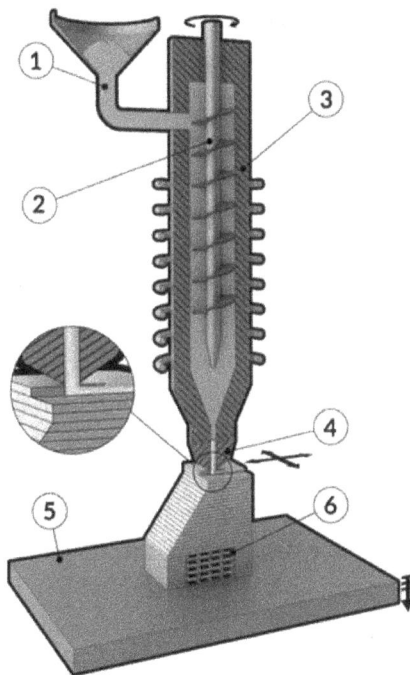

Fig. 4.21   Extruder for granulate processing.

Source: R.J. Crawford Plastics Engineering (Third Edition), 1998, *Processing of Plastics*.

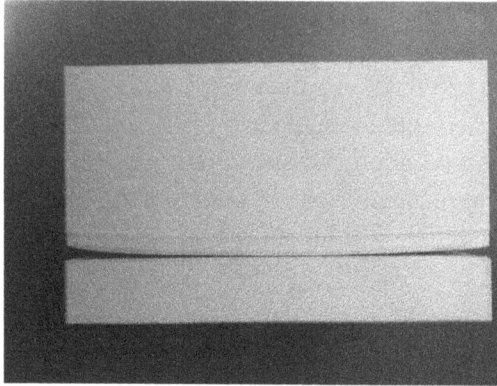

Fig. 4.22   Warping causing delamination in 3D printing of plastics.

Source: Authors.

up down moving stamp and expelled at the bottom by a nozzle (piston type extruder). However, the material flow to the workpiece varies then with time, so that the applied traces vary both in width and height. This ripple could be compensated by the fact that the same nozzle is fed by a second cylinder, which is compressed and relaxed in the opposite way to the first cylinder (twin piston).

Another problem is the shrinkage of the hotly applied material during cooling, because it distorts the required dimensions (Fig. 4.22).

Solutions include oversized printing and subsequent milling to achieve the exact dimensions, but this eliminates certain advantages of 3D printing, such as avoiding post-processing. Another and elegant possibility is the continuous temperature measurement in the current processing point and the computational adjustment (increase) of the material mass extruded per unit length of the travel path.

The latter problem of the distortion of the components produced in 3D printing, the so-called warping, has to be dealt with in detail. This is due to the fact that the last applied layer, which is still considerably warmer than the layer under it, shrinks during cooling and thus creates stresses, in particular compressive forces parallel to the traversing direction, which lead to a bulge of this layer and possibly leads to a delamination, a lift-off from the rest of the workpiece or from the construction platform. This effect becomes weaker the smaller the temperature difference between the individual layers is, as both then shrink to the same extent when cooled down. The material also plays a role, whereby ABS (hard thermoplast), unlike PLA (polyester thermoplast), is much more inclined to warping. This effect has a particularly strong effect on long components, since with increasing length also the shrinkage and thus the length difference of the top and the underlying material increases, so that for the intended length compensation larger forces are necessary, which lead as compressive stresses to stronger bulge of the lower fiber. This effect can be combated by successive longitudinal interruptions, such as holes or slots, as already mentioned above. A fiberglass content in the material causes antiwarping as the shrinkage is reduced to zero, what is possible with the use of granules independent of available filament.

Lower layer thicknesses also reduce the tendency to warping, as a slender buckling rod with which one can compare the underlying layer requires less force for deformation.

Mention was made of the heating of the construction platform and the entire building space, which was helpful for avoiding delamination, both of which were not trivial in view of the large volume. As far as the latter heating is concerned, heated air from the outside can pass through the building space in the longitudinal direction or selective heating of the work piece by infrared emitter or recirculation heating, while the construction platform will have to be heated by inserted heating wires or inductively.

References

1. Based on Chapter 2, Materials and workpiece classification, by Thomas Varga and N. Spörk. In Handbook of the Eurolaser Academy, Editor D. Schuöcker, Chapman and Hill, 1998.
2. Beyer, Eckhard, Schweißen mit Laser Grundlagen, Herausgeber: Herziger, Gerd, Weber, Horst (Hrsg.) Springer, 1995.
3. U. Dilthey, Schweißtechnische Fertigungsverfahren 1, 3. Aufl., Springer, Heidelberg 2006 …
4. M. Resch, Alexander FH Kaplan, Dieter Schuoecker, "Laser-assisted generating of three-dimensional parts by the blown powder process," Proc. SPIE 4184, XIII International Symposium on Gas Flow and Chemical Lasers and High-Power Laser Conference, (25 January 2001);
5. Generieren von 3D Teilen aus Metalldrähten mittels Laserstrahlung Austrian patent 515465.

Chapter 5

Laser Processing without Mass Change

5.1 Overview

Laser hardening can be performed if a laser beam scans over the surface of a steel workpiece with speed and power sufficient to reach a temperature somewhat above 700°C. After the beam has moved on, heat conduction into the bulk of the material leads to rapid cooling resulting in strong hardening at and near the workpiece surface.

Laser reinforcement can be obtained if a few narrow laser hardened stripes are generated at the surface of a metal sheet that is stiffened and bulging can be avoided.

Laser polishing is performed by moving a laser beam with appropriate power and speed zig zag over a workpiece thus melting a thin surface layer. Due to surface tension the melt cancels out all the irregularities of roughness, mainly peaks, and thus the melt surface becomes smooth also after the solidification, resulting in reduced roughness.

Laser forming is obtained if a laser beam moves across a metal sheet and heats it selectively resulting in thermal expansion of the material in a narrow zone scanned by the laser. The latter expansion of the hot region exerts pressure to the cooler surrounding that leads to a buckling of the workpiece. The latter effect can be used for bending.

Laser assisted forming is obtained if a laser beam heats selectively those regions of a workpiece to be formed where strong deformations are expected. The elevated temperature in these regions leads to reduced forces necessary for plastic deformations and also the rise of the maximum expansion at rupture that limits the degree of deformations. Of course these applications are restricted to thin sheet metals since laser heating cannot penetrate more than roughly a millimeter into the material.

Options are **die bending** especially useful for brittle materials, **inline profiling**, **deep drawing** and **hydroforming**, all processes for sheet metals where strong deformations take place in small regions well suited for laser heating. The same is true for **wire drawing** that is usually performed with narrow nozzles that suffer from wear but can be omitted due to laser heating.

5.2 Laser Hardening

If *ferritic* steel with 0.8% carbon is heated to a temperature of around 730°C — the so-called transition point — an *austenitic* structure arises in which the Carbon atoms are regularly dissolved and arranged at interstitial sites. However, if the steel is now cooled rapidly, the structure changes again and the carbon atoms have to leave their site, but as a result of the rapid cooling, they do not have enough time to arrange themselves and remain irregularly, thus forming defects in the lattice, that are associated to internal stress and a very high hardness appears (details see Chapter 10). The latter microstructure is called a *martensitic* structure.

This hardening process can now be done very well with a high power laser beam, which is moved across the workpiece and achieves the heating to the transformation point at every point.

After the laser beam has moved on, the points left by the beam are cooled rapidly by heat conduction into the interior of the workpiece, so-called thermal quenching, takes place and thus a very hard martensitic structure forms. If several such traces are now placed shoulder to shoulder, the whole workpiece surface and a thin layer below are hardened. If the laser beam hits during this movement workpiece steps, protrusions or holes, the energy balance is disturbed, whereby usually an increase in the workpiece temperature takes place as a result of improved absorption and reduced heat conduction. However, this overheating leads to erratic melting of the workpiece that disturbs after resolidification the surface geometry and thus after hardening, post-processing is necessary and costs additional time and money.

In order to avoid the latter effect, a process has now been developed in which the temperature of the workpiece in the point of incidence of the laser beam is measured and used to control the laser power, thus ensuring a constant temperature at the processing point, regardless of its spatial structure, with holes, steps or protrusions.

The actual temperature-controlled hardening system (Figs. 5.1 and 5.2) therefore consists of a 6-kW disc laser, a glass fiber that guides the laser radiation to the machining head, a second light conducting fiber that collects the thermal radiation going out from the hot focus of the beam on the workpiece and conducts it to a pyrometer, and a unit that controls the laser power by the measured temperature, thus keeping the latter constant.

Fig. 5.1 Laser hardening stand with 6 kW disc laser (left), gantry robot with scanner and hardening system (middle) and video observation of the laser focus on the workpiece (right).

Source: Authors.

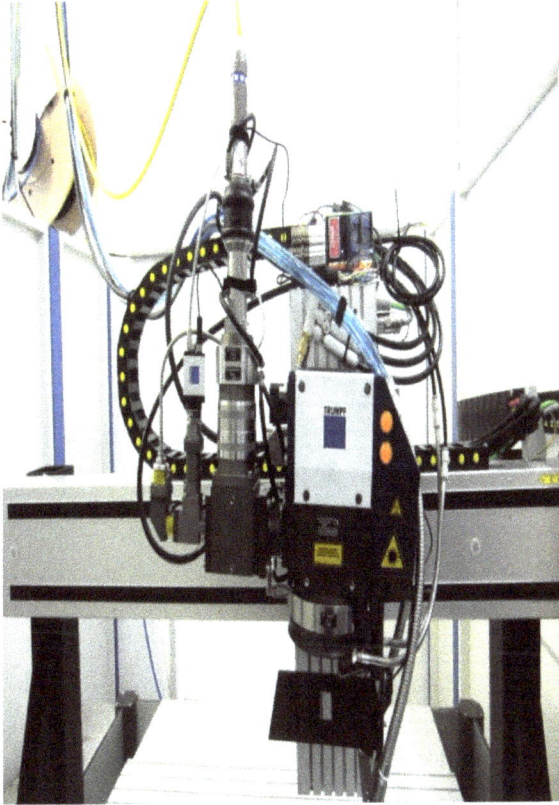

Fig. 5.2 Laser hardening head with observation optics, fiber coupled pyrometer for the measurement of the focus temperature and focusing optics coupled by a second fiber to the hardening laser (LC Gmunden, 2013).

Source: Authors.

A portal robot with scanner, which on the one hand swings the laser beam transversely to the processing direction over the workpiece in order to achieve a hardening track up to 30 mm wide and, on the other hand, moves it in the direction of processing completes the whole arrangement. This function is not contained in the standard control software of the scanner and can only be realized if one of the scanning mirrors is directly controlled by a triangular voltage signal. So for the control of the scanner mirrors a new auxiliary control system had to be developed. During the

Fig. 5.3   Construction steel hardened with the RHCU system avoiding melting at the edges of the track.

Source: Authors.

"wobbling" process, the control module also has to reduce the laser power near the edges of the hardened track in order to avoid overheating of the workpiece. An example of the correct function of the latter facility is shown in Fig. 5.3.

That no melting actually occurred when the laser reached surface distortions shown in Fig. 5.4, in the case of a workpiece made of structural steel hardened with the new process and proves thus the post-processing freedom of this new laser hardening process.

5.3 Reinforcement by Laser Hardening

Thin sheets, such as those used for car bodies, suffer from deformations, as bumps, if point-shaped forces, as in the case of stone fall, act on them, which cannot be tolerated. The depth of the bump depends then on the resistance of the workpiece against elastic and plastic deformation.

The latter is usually increased if kinks are inserted into the sheets or also slats are attached. Both measures require additional time, material and energy expenditure in the production of sheet metal parts. Conserving

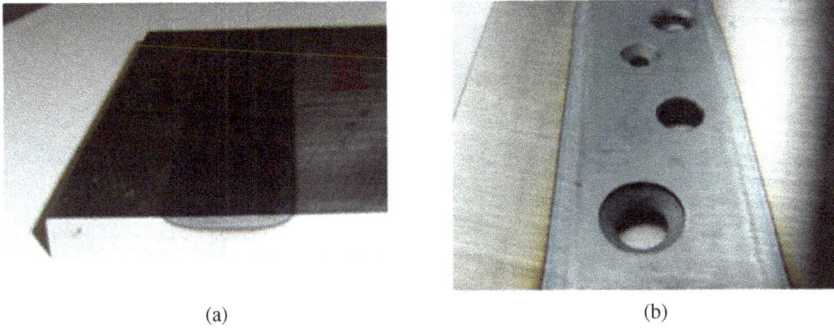

(a)   (b)

Fig. 5.4 Laser hardened strip on construction steel (3 mm thick), (a) cross-section with 700 HV; (b) Strip with diverse holes, no surface defects due to melting (Gmunden, 2013). *Source*: Authors.

resources can be obtained if stiffening is carried out by introducing narrow traces of hardness created by lasers, which has already proved feasible in preliminary tests (Figs. 5.5 and 5.6).

The effectiveness of the hardness traces with regard to increasing the bumping stiffness is indicated by Fig. 5.6 whereby sheets made of TRIP steel without and with hardness traces are applied with a constant point-shaped force and the resulting permanent plastic deformation, i.e., bulge depth (stamp path), has been measured.

It was shown that with two cross-shaped hardness traces a maximum improvement is achieved compared to workpieces without hardening, but this, if even more traces are produced, is reduced, which depends on the fact that then the first produced traces are softened again.

5.4 Laser Polishing

Laser polishing developed by the Fraunhofer Institute of laser technology, Aachen, can be performed by melting the surface of a rough workpiece up to a depth of 100 micrometer by CW or pulsed laser with relatively low power <100 W and with pulse durations of 100 ns. The melt then gets smoother due to surface tension that cancels peaks and other protrusions causing roughness. With this process from initially some micrometers a roughness in the order of magnitude 0.1 micrometer can be reached at a

Fig. 5.5   TRIP steel sheet with hardening traces (below) and setup for measuring permanent bulging performance.

Source: Authors.

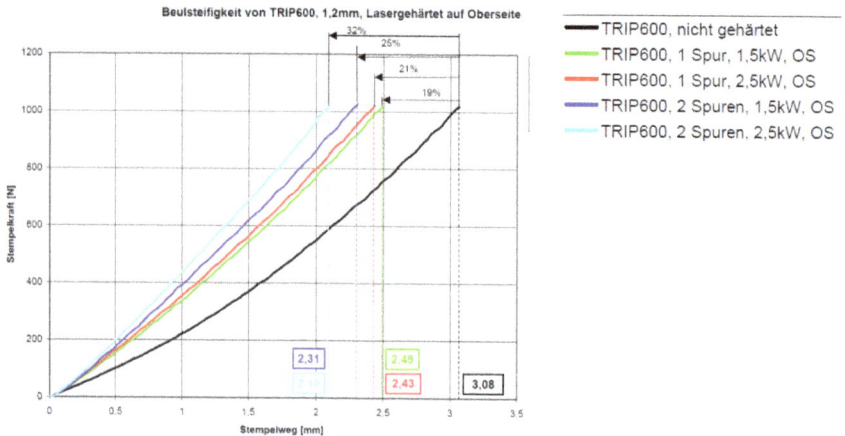

Fig. 5.6   Permanent deformation of TRIP steel due to embedding of hardened traces dependent on number of traces and laser power (1 no hardening, 2 one trace, 1.5 kW, 3 one trace 2.5 kW, 4 two traces, 1.5 kW, 5 two traces, 2.5 kW).

Fig. 5.7 Laser polished part. Von Bestemrc — Eigenes Werk, CC BY-SA 4.0, https://commons.wikimedia.org/w/index.php?curid=94200072.

rate of below 5 cm^2/min. Many metals, as steels and nickel or titanium alloys can be treated successfully.

5.5 Laser Forming

5.5.1 *Basic principle*

If a strip of sheet metal is subject to opposite forces acting on both ends, it is elongated and in a perpendicular direction constricted, since no mass change is possible. Initially the deformations vanish if the forces are removed, since the effect of the forces was only a displacement of atoms in the lattice without breaking the bondage of the atoms. This *elastic deformation* changes to a permanent, *plastic deformation* if the *yield strength* is reached since then the forces are strong enough to break out atoms from their position in the lattice and to settle them on new ones resulting in the displacement of mass and thus permanent elongation of the metal part that is maintained even after the forces are removed. With rising forces the elongation rises until rupture takes place at the tensile strength (see Chapter 8).

It is very important for the application of lasers in metal forming that all three quantities limiting the plastic regime, namely yield and *tensile strength* and elongation at rupture, the *ductility*, depend strongly on the temperature, the two strengths decreasing and the ductility rising with rising temperature since due to the enhanced movement of the atoms the bonds between the atoms become weaker. As an example for steel the yield strength decreases by some 10% and the ductility rises by a multitude of 10% dependent on the composition if the temperature reached 800°C.

The latter behavior allows to facilitate forming and avoid breaks in forming sheet metal parts with strong deformations as in bending, profile rolling, hydroforming and wire drawing as well as deep drawing since the restricted regions of sharp deformations and the low thickness allow heating with a laser beam. The latter applications are treated below. All these processes have been experimentally investigated and their feasibility has been proved, but practically none of them has reached industrial use up to now.

5.5.2 *Laser bending*

Below the transformation temperature, for example from steel, there is no change in the structure, but heat stresses are built up. These heat stresses can be used for deformation, for example for bending a workpiece. In this process a focused laser beam travels along the desired bending edge over the workpiece, that it is heated on the surface, thus expanding and forming a bend convex towards the laser beam. Subsequently, the heat flows to the bottom of the workpiece, so that it constricts again on the surface where the laser was heating and expands at the bottom, thus reversing the bend to concave with respect to the laser beam. If the plastic regime is reached, a permanent bend remains. With one passage of the laser beam, however, only a bending angle of a few degrees can be realized for steel and aluminum (see Fig. 5.8).

5.5.3 *Laser assisted bending*

If a thin piece of sheet metal is bent, the outside surface is expanded and the innermost surface compressed. The extension of the expanded region is determined according to Fig. 5.9 by the average radius of the bend that must always be larger than zero even if the inner surface has a sharp kink instead of a round shape. The maximum bending is reached if the above

Fig. 5.8 Laser bending.

Source: Bachmann, Dickey & Lazarus, Making light work of metal bending: Laser Forming in Rapid prototyping, Quantum beam science, USA 2020.

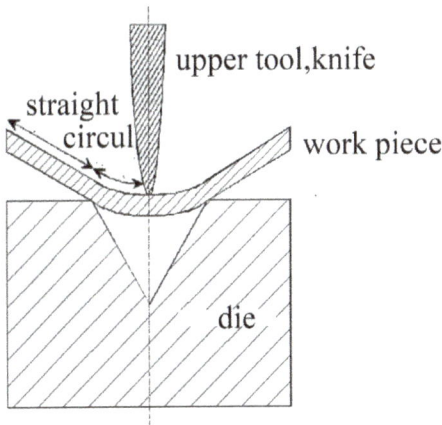

Fig. 5.9 Die bending.

Source: D. Schuöcker, Spanlose Fertigung, Fig. 4.3.1 Oldenbourg, München 2004.

mentioned expansion approaches the ductility, the expansion at rupture. Since the latter quantity rises with heating as mentioned above, elevated temperatures at the bending edge and its vicinity allows stronger bending especially for brittle materials. Moreover the bending force is reduced. Since the workpiece is thin and the region with deformations is small due

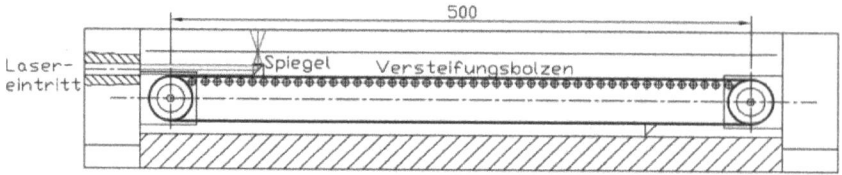

Fig. 5.10 Paternoster arrangement to heat a workpiece along a line.
Source: Authors.

Fig. 5.11 Experimental set up for a paternoster mounted on a bending die.
Source: Authors.

to its dependence on the small inner radius of the workpiece selective laser heating is an optimum choice leading to *laser assisted bending* [1].

Laser heating is necessary only in a narrow line shaped region along the bending edge and can be realized in various ways: First a focused beam can be moved along the bending edge forth and back. A practical solution is the *Paternoster* (see Figs. 5.10 and 5.11) with a mirror reflecting the beam of a solid state laser with an incidence parallel to the bending edge and reflecting it to the workpiece while moving along the bending line. The latter device can be integrated in various types of bending machines.

A most recent solution is the use of diode lasers that are arranged as an array in parallel and along the bending edge (Fig. 5.10). Other solutions

Fig. 5.12 (Left) Diode laser array 10 mm wide with 200 W beam power, (right) 8 arrays mounted in in a box 100 mm wide with beam power 1600 W and massive current leads. Arrays are water cooled.

Source: Authors.

Fig. 5.13 4 bending dies, 2 opened, 1 showing laser box with current and water connectors fitting in the next (missing) die.

Source: Authors.

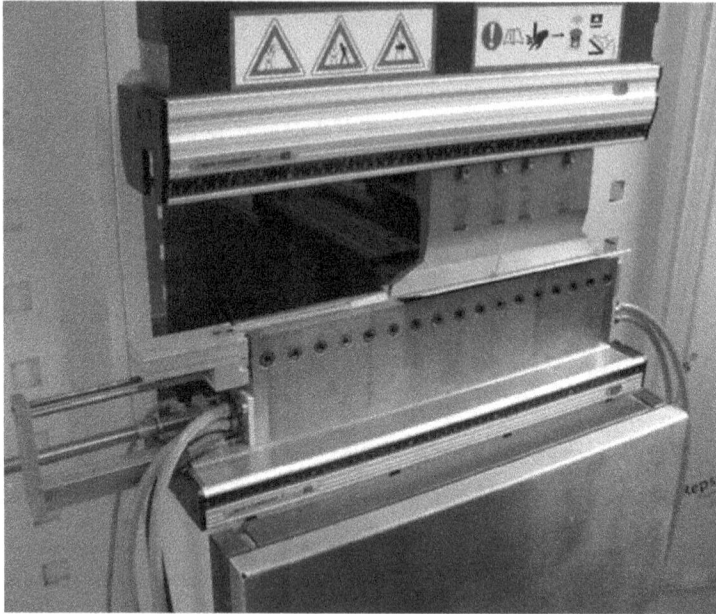

Fig. 5.14 Die bending machine with 4 diode armed dies just acting on a piece of steel sheet.

Source: Authors.

use rotating segmented mirrors or stationary beam forming optics but do not reach the quality of heating as provided by the above systems.

In detail the Paternoster uses an endless rubber belt moved by two rotating rolls, one of them motor driven. The latter belt carries two mirrors as mentioned above where always one of them is on the upper side of the belt and heats the workpiece while the other one is below and inactive. The belt moves the active mirror along the bending edge and leaves the upper side while the second mirror enters the upper side thus heating the workpiece continuously. Laser radiation is fed to the arrangement by a fiber coupled solid state layer of recent design that ensures a low divergence that keeps the beam collimated throughout its way along the bending edge to the reflecting mirror. Under the latter condition the Paternoster can be energized with nearly unlimited beam power and is thus suited for thick workpieces.

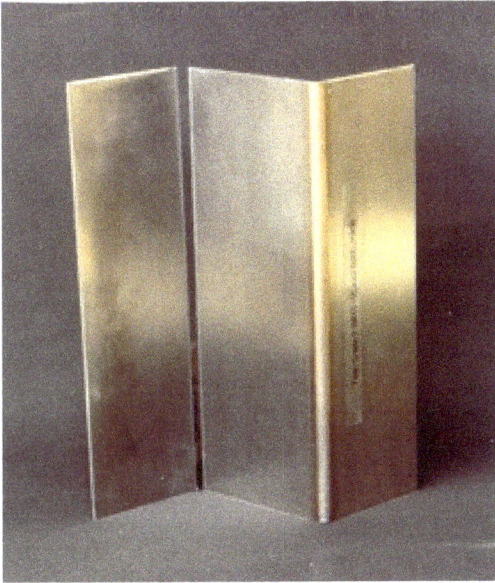

Fig. 5.15 Bending of Titanium with and without laser heating.

Source: Authors.

The latter device can be mounted on one cheek of a bending die, what means that the workpiece must first be heated and then shifted to the die where bending takes place.

The disadvantage of the paternoster can be avoided if a linear array of laser diodes as mentioned above is used. The latter are contained in a box that fits into a bending die. A complex arrangement from lenses, mirrors and prisms ensures that out of the many single beams a light plane perpendicular to the workpiece is generated well suited to heat the bending edge. The diode box is fitted with connectors on both sides that allows to supply electrical current and cooling water and to hand it over to the next box if several dies are arranged in series to bend wide parts.

A numerical evaluation of laser assisted bending for 4 mm magnesium 40×100 mm^2 with 5s irradiation yielded the diagram Fig. 5.16 with bending force vs. stamp way for some values of laser power. The end of the curves indicates rupture. The diagram clealy demonstrates the effects

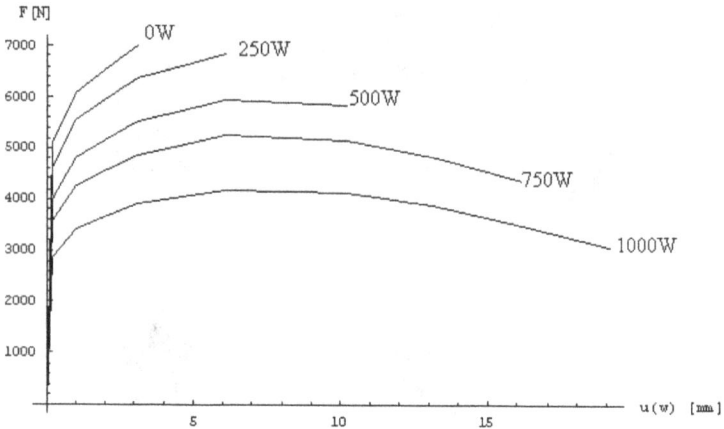

Fig. 5.16 Bending force vs. stamp way dependent on laser power.
Source: Authors.

of laser heating: First, the force necessary for bending is reduced and second, bending can become stronger without a risk of cracks or rupture.

5.5.4 *Laser assisted inline profiling*

Very similar to die bending is inline profiling, that it is used to produce endless profiles of steel, aluminum and other materials. For this process roller pairs are used, whose surface is not cylindrical, but have a diameter varying along the roller axis, thus enforcing a certain lateral profile to the endless metal strip (Figs. 5.17–5.19).

If now the profile contains sharp bends that can give rise to cracks or even breaks, laser heating prior to rolling at those rolls where kinks are formed, that means just before the strip enters the pair of rolls, can improve the situation by an increased ductility quite similar to laser assisted die bending. Practical experiments at TU Wien yielded the dependence of pressure on the upper roll vs. way down of the roll without laser and with laser (Fig. 5.20). It points out as expected that stronger bending of the rolled material can be obtained with laser heating.

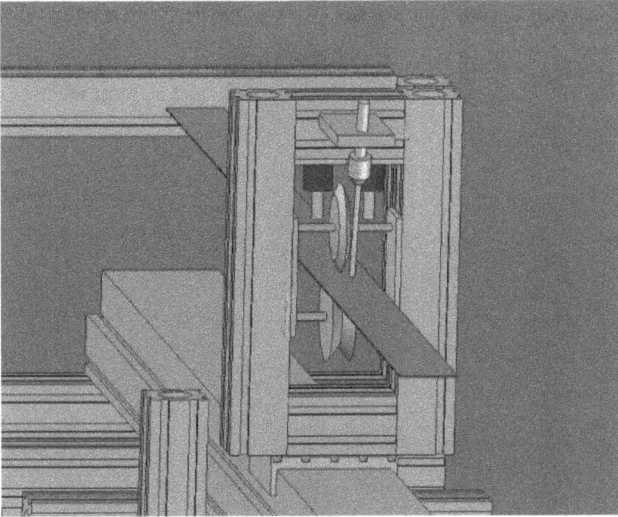

Fig. 5.17 Inline profiling with one pair of profiled rolls, also showing laser beam to support profiling.

Source: Authors.

Fig. 5.18 Experimental setup for laser assisted inline profiling.

Source: Authors.

Fig. 5.19   Pair of profiling rolls, separated during adjustment and 1 kW diode laser.
Source: Authors.

5.5.5 *Laser assisted deep drawing*

Deep drawing (Fig. 5.21) is a sheet metal forming process that uses a stamp with a shape that fits precisely into a die where the geometry of the first one corresponds to that of the second but leaves a certain allowance to take the workpiece. The stamp presses the workpiece into the die and the initially flat material is bent at the edge of the die and is drawn into the die. Thus the sheet assumes the shape of the latter. To prevent take off of the remaining flat material during drawing that would cause twinkles in the finished workpiece a *blank holder* is used. Normally this process is used for small part as pots and other kitchen suppliants but also for very large ones as car body parts. The latter usually are reinforced by beads and

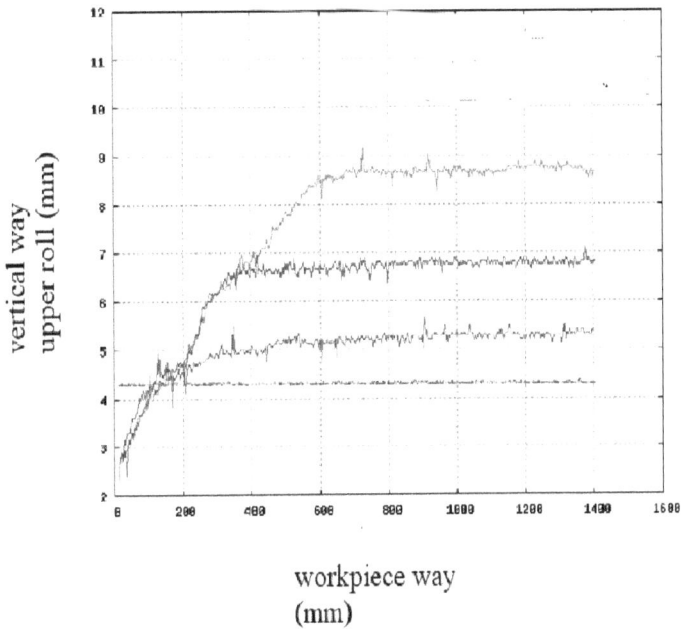

Fig. 5.20 Vertical distance travelled by the upper roll during the movement of the rolls with respect to the workpiece without laser heating (bottom), heating with a laser power of roughly 600 W (middle) and with 1 kW (top).

Source: Authors.

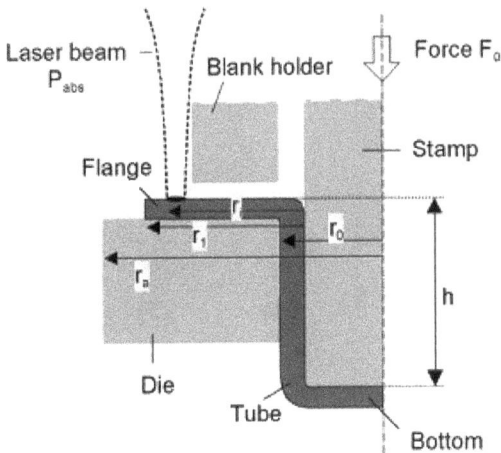

Fig. 5.21 Deep drawing with three tools and laser assistance.

Source: D. Schuöcker, Spanlose Fertigung, Fig. 4.4.1 Oldenbourg, München 2004.

kinks and contain sharp bends, quite similar to bending. The latter are associated to a threat of cracks or rupture that can be avoided quite similar to laser assisted bending with laser heating that is very well suited since only restricted areas of thin material have to be selectively softened to facilitate strong forming actions.

The simplest deep drawing part is a cylindrical pot as shown above. The latter contains a flat bottom, a vertical tube and as long as it is not fully completed a *flange* that vanishes when drawing is complete and all the initially flat material is now settled in the pot. Sharp bends are located at the edge of the opening in the die, where heating can be performed easily by moving a laser beam around the frame.

Drawing of little pots from steel without and with laser assistance has been experimentally investigated at TU Vienna and showed that with parameter sets as workpiece thickness, geometry of the die and drawing force, where rupture at the ground of the pot happens, the latter can be avoided by sufficient laser heating (Fig. 5.22). These experiments were done by using a hydraulic drawing press (Fig. 5.23). The workpiece was

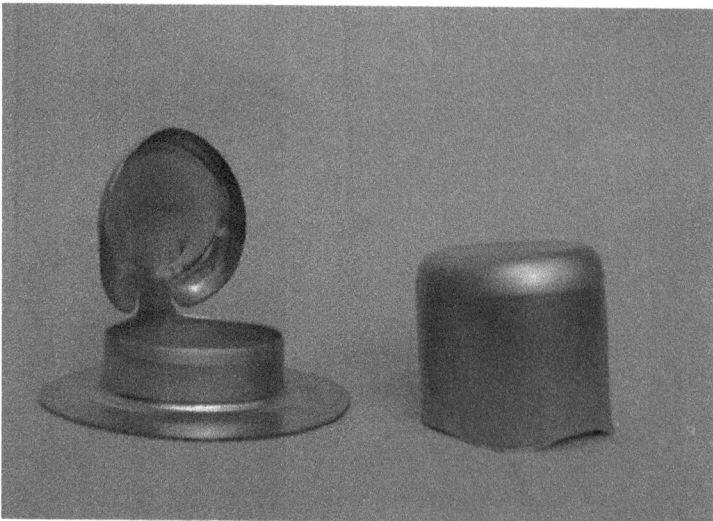

Fig. 5.22   Steel pot drawn without (left) and with (right) laser heating.
Source: Authors.

Fig. 5.23   Deep drawing press with laser heating outside the press.
Source: Authors.

heated outside the press with a laser beam moved across the latter and finally shifted into the press.

The latter solution is practically unfeasible and therefore further experiments have been made with a long and narrow finger that conducted the laser radiation to its end. An optical assembly focused the beam and directed it towards the work piece already located at the die and waiting for drawing, being heated by the finger and finally drawn by the stamp (Fig. 5.24).

The latter solution is much more feasible, but nowadays as high performance diode lasers are available, it seems to be best to integrate a

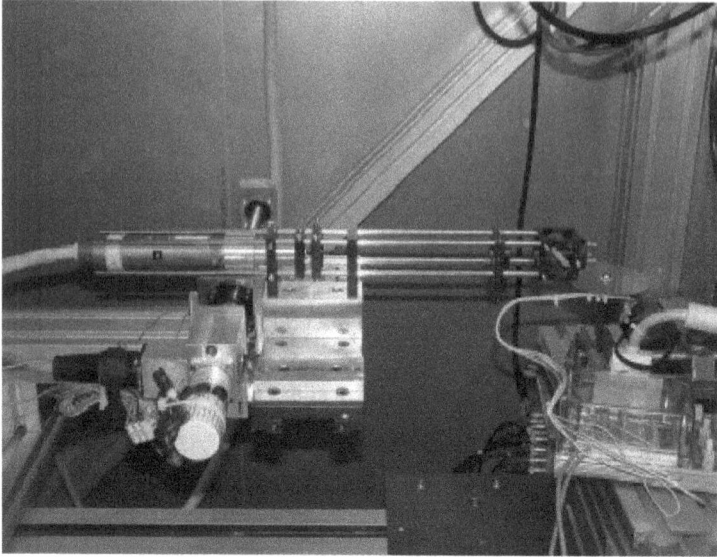

Fig. 5.24 Deep drawing with laser heating by a finger that carries the beam guiding fiber plus the focusing optics.

Source: Authors.

multitude of these lasers directly in the die or the stamp, where strong forming actions will take place.

5.5.6 *Laser assisted hydroforming*

Internal high-pressure forming (IHU) is a process in which a hollow blank is first inserted into a two-piece die, the inner shape of which corresponds to the shape of the workpiece to be manufactured. Water is then discharged into the blank at very high pressure, whereby the latter attaches itself to the recess of the die and thus takes its form.

An example of such a forming is the production of a T piece from a cylindrical tube (Fig. 5.25).

The required strong forming leads to a corresponding hardening, which then prevents a further transformation which is necessary to achieve

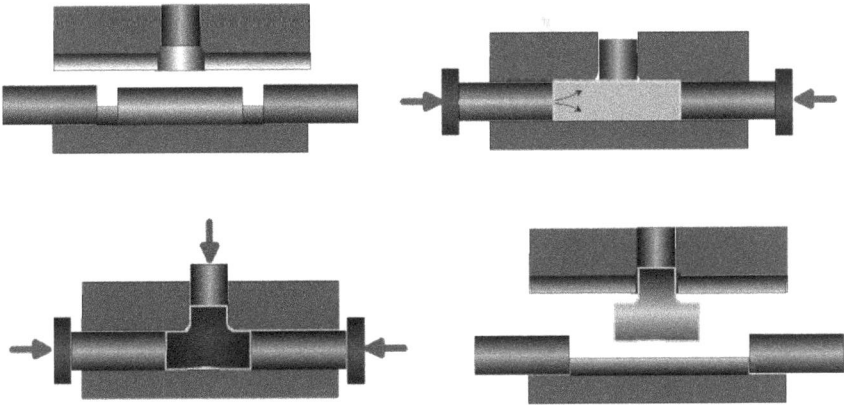

Fig. 5.25 Sequence of hydroforming, (upper left) putting the initially cylindrical part in the tools, (upper right) filling with pressure water, (bottom left) formation of a "dome", (bottom right) taking the finished part out of the tool.

Source: Authors.

the final form. So far, this problem has been solved by the fact that between two forming steps the workpiece was *annealed* (see Chapter 9.) in a furnace, which consumes time and energy. Since strong forming occurs only in limited zones of the workpiece, which also applies to hardening, selective laser heating (at 1000°C) can save time and energy.

A practical example for a work piece hardened during the first forming steps, as bending and then hydroforming the dome and thus work hardened in and near the dome shows Fig. 5.26. If then the last forming step, sharpening of the edges of the dome, is carried out, the workpiece experiences cracks, since the ductility reduced due to work hardening does not allow the necessary strong elongation of the material on the outer surface of the workpiece. A work piece with the latter cracks is shown by Fig. 5.27.

In order to avoid the above cracks, annealing of the work piece in a furnace as mentioned above is absolutely necessary. A better solution is offered by laser assistance that allows a selective annealing only in those regions of the work piece, where strong deformations have been achieved and thus work hardening has been obtained.

Fig. 5.26   Typical hydroforming part after the first two forming steps.

Source: Authors.

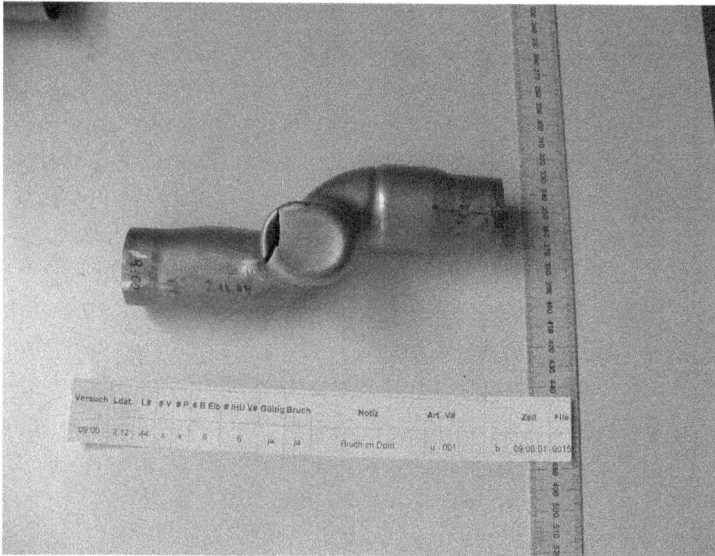

Fig. 5.27   Hydroforming part after the third and final forming step.

Source: Authors.

Fig. 5.28   Processing head for laser annealing with incorporated temperature acquisition, (Right) Fiber supplying 3 kW YAG laser radiation for heating, (left) Fiber delivering radiation of the heated workpiece to a remote pyrometer (TU Wien, 2007).

Source: Authors.

In order to carry out an experimental feasibility study an apparatus has been build up that uses a 5-axes movement between the workpiece and the laser beam to scan the desired part of the surface of the work piece with a high power laser beam of a Nd:YAG laser connected to the experimental set up by an optical fiber. Since for the purpose of a removal of work hardening the temperature must remain within narrow limits above 1000°C, a temperature control has been accomplished. The latter consists mainly of a pyrometer that is connected to an optical fiber collecting thermal radiation of the laser heated region and uses a fast micro processor to control the laser beam power according to the measured temperature of the work piece.

Figure 5.28 shows the optical systems for heating of the work piece and for the measurement of the work piece temperature.

With the latter system heating of the workpiece prior to the last processing step in order to compensate work hardening by annealing has been carried out for the work piece shown in Fig. 5.26 thus avoiding the time

Fig. 5.29   Finished workpiece of Fig. 5.26 after the first two forming steps and selectively annealed with the laser system.

Source: Authors.

consuming heating in an oven. Figure 5.29 shows than the workpiece of Fig. 5.26 after the last forming step. It clearly points out that the workpiece shows at the roof of the so called "dome" very sharp edges but nevertheless remains without any cracks, thus proving the feasibility of laser assisted hydroforming.

In this context it should be mentioned that a similar application of selective laser annealing was also successfully carried out at the laser center in Gmunden on behalf of a leading laser manufacturer, even in already hardened work piece areas in which processing steps such as the production of recess and others have to be carried out.

5.5.7 *Die-less wire drawing with laser heating*

Thin wires with a diameter below 2 mm and made of ferrous or refractory materials usually are produced by drawing the wire through subsequent dies with decreasing hole diameter. Besides the disadvantage of strong wear of the dies due to friction, for the same reason scratches are caused on the wire surface that may reduce its strength. Moreover, in the case of brittle materials as highly alloyed and of high strength, corrosion resistant steels or refractory metals as tungsten, that are generally difficult to form, rupture is likely. It should also be mentioned that with one die only a limited reduction of the wire diameter can be obtained due to work hardening. Therefore, in order to obtain a considerable diameter reduction, a sequence of dies must be used where heating of the wire after each die reduces the effects of work hardening. Of course the problem of work hardening also applies to less brittle and ductile materials such as ordinary steels and others.

The latter problems can be solved by using die-less drawing processes where the wire is heated in order to reduce the yield strength to a value clearly below the axial stress exerted on the wire and thus allowing constriction and elongation of the wire. To avoid wire breaks cooling must be carried out shortly after the heating of the wire in order to raise the yield strength again above the applied tensile force.

Compared to other heat sources like induction or resistance heating, heating by laser radiation has several advantages. Especially for fine wires inductive heating as well as resistance heating are difficult to control whereas the beam of a laser offers high flexibility and can be manipulated easily.

Other advantages are: The temperatures can be controlled by changing the laser output power very fast, wire diameters can be changed easily, large diameter reductions can be achieved in a single step, drawing of difficult materials is possible, thermo-mechanical treatment is possible, no lubricants are required, varying diameters are achievable.

Nevertheless, up to now, process stability or an unwanted and uncontrollable variation of the wire diameter was one of the major problems in laser assisted wire drawing.

To investigate the feasibility of die-less wire drawing with laser heating the Institute for Forming and High Power Laser Technology of the University of Technology in Vienna built an experimental setup that contains a 1 kW diode laser for heating of the wire and a gas flow unit for

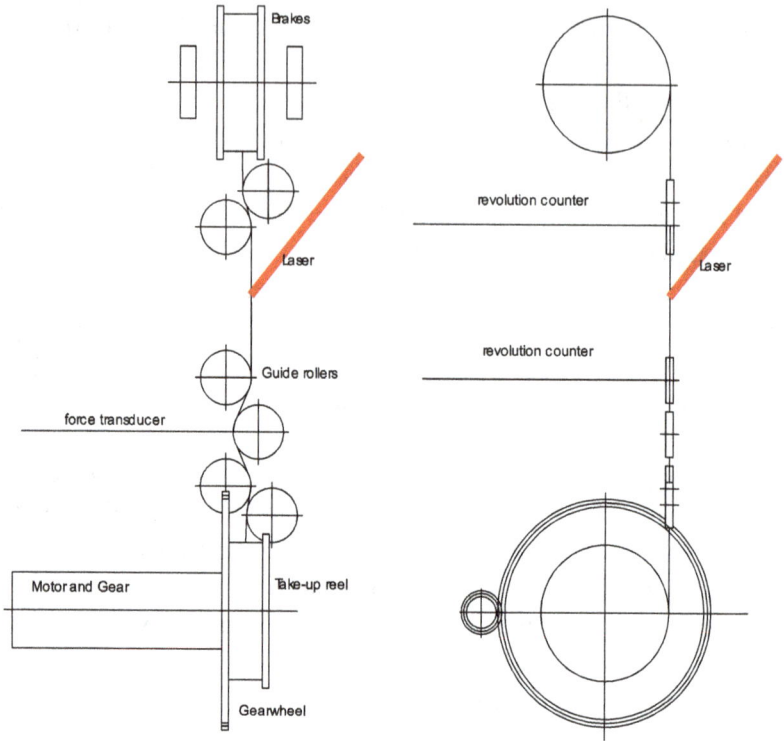

Fig. 5.30 Schematic view of laser heated wire drawing system, left: side view, right: front view.

Source: Authors.

cooling (Figs. 5.30 and 5.31). Figure 5.30 shows the schematic layout of the laser drawing system and Fig. 5.31 a photograph of the practical realization. Figure 5.32 demonstrates the performance of the system in terms of reduction of wire cross section or steel and copper and Fig. 5.33 showes the dependence of wire speed vs. laser power.

Fig. 5.31   Laboratory setup for die-less wire drawing with laser heating.

Source: Authors.

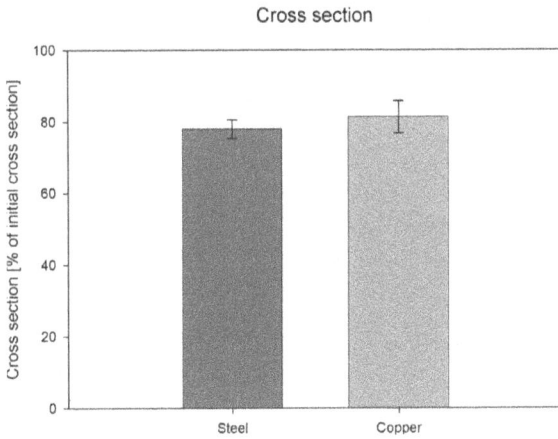

Fig. 5.32   Reduction of then cross section after one pass for steel (left) and copper (right).

Source: Authors.

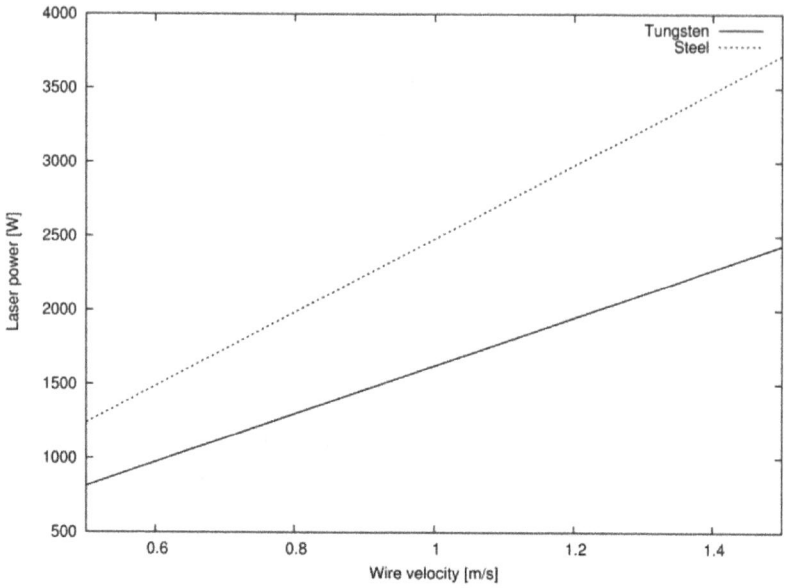

Fig. 5.33   Wire speed vs. laser power.

Source: Authors.

Reference

1.   F Kilian, K Schröder, D Schuöcker — US Patent 6,415,639, 2002 — Google Patents.

Chapter 6

Laser Safety

6.1 Introduction

Light beams generated by lasers usually show a small cross section in the order of some mm to some cm. Thus even with relatively low beam power relatively high intensities are achieved, that can lead to an overheating of human tissue exposed to the laser radiation as the eye or the skin, even if the radiation is unfocused. More serious harm can of course be done, if a focused beam hits a human body. Moreover the mere burning through laser radiation turns to much more severe effects, if the wavelength of the radiation is in a certain range of the ultraviolet region, where even cancer can be caused. Nevertheless besides these dangerous properties of laser radiation, the protection against the latter harms is facilitated by the directed propagation of a laser beam, that extends in a cone with a very small opening angle, the divergence, thus limiting the interaction with a laser beam to a restricted volume, that can be mantled with a solid protection wall, thus preventing the beam to heat portions of human bodies. Moreover due to the very small frequency bandwidth of laser light it is possible to protect the eye against it by narrow band filters, that can easily be realized. Besides the hazards of laser radiation some lasers, especially high power lasers, are dangerous in various ways, as for instance due to the use of poisonous and aggressive gases, as in the case of excimer lasers, due to the use of high voltages, and due to high power-high frequency currents in the case of modern carbon dioxide lasers.

Moreover, in material processing the application of lasers can be dangerous due to uncontrolled reflection of radiation and due to the deliberation of matter in various dangerous states as hot molten droplets or sparks, carcinogenic dust or poisonous gas, for instance in the case of cutting plastics. Besides these hazards, material processing sometimes also leads to the generation of shock waves. In addition to the safety measures with respect to the laser beam mentioned above, safety precautions against all these dangers associated with the laser source and the laser application must be provided, as for instance appropriate housings of the source or the whole laser system.

So at the end, the use of lasers in various applications is by no means more dangerous than conventional mechanical tools as for instance a saw blade, that can cause most serious harm, if it bursts and the parts explode with high speed. Nevertheless, the user of lasers must be aware of all dangers and hazards of lasers and their application. To be able to protect oneself against the hazards associated with lasers, especially of those with high power, one must know of all hazards of laser sources, beams and applications. So the majority of this chapter is devoted to the main hazards associated with laser beams and their applications as well as safety precautions and measures, that can safely prevent accidents.

6.2 Hazards of Laser Technology and Safety Precautions

6.2.1 *Hazards of high power laser sources*

Obviously pumping of the active medium, that is carried out in the case of high power lasers for materials processing with considerably high powered energy can be a source of hazards, whereas in the case of CO_2 lasers either DC-voltages with a magnitude of several 10.000 V or short-time pulsed voltages with an even much higher magnitude (also in excimer lasers) or even high power RF-currents are used. It is less obvious, that also RF-currents can become quite harmful, although they are usually conducted along the skin and cannot penetrate into the depth of the human body due to the well known skin effect at high frequencies. Nevertheless, freely propagating RF waves, that are of the same nature as light, but have a much larger wavelength, can be absorbed by human tissue and heat it up, whereas several organs show a strong sensibility against overheating of a few centigrade. One example are for instance the roots of the teeth, a second example are the male testicles. In the case of pulsed voltages for carbon dioxide and excimer lasers a special threat are the large capacitors, that are used to store electrical energy, that is then deliberated in the form of a short and powerful pulse. In order to prevent harmful actions of high voltages, all parts, that are connected to high voltages, must be safely mantled with a shielding material which is securely grounded. Moreover, any removal of shielding parts must activate the so-called 'interlock switches, leading to an immediate interruption of the power supply to the laser. In addition all capacitors, for instance used to smoothen rectified currents, must be safely discharged on power-off. Special attention must of course be paid to capacitors used for the storage of pulse energy as mentioned above, as they have not only to be discharged, but also — to be on the safe side — must be short circuited and grounded. So far, in the case of DC and pulsed voltages the prevention of a direct contact between parts under high voltage and the human body is necessary. In the case of RF waves, which appear in modern high power carbon dioxide lasers, since the latter are not excited by DC currents but rather by RF energy with its advantages of elevated power-to-volume ratio, the emission going out from the laser must be avoided. Therefore a so-called "Faraday cage" formed by an electrically conducting grid or net, that covers all parts that could emit RF radiation and which is safely grounded, can totally prevent harmful RF radiation. In practice instead of a grid or net also sheet metal parts with holes, that allow to look inside the laser and that also reduce the weight, are used. To be totally safe, it is advisable to include also some sensors, that indicate the presence of an RF field, in the arrangement of the cage mentioned above, if for instance for any reason the shielding of the Faraday cage must be removed. Nevertheless, this shielding must also be protected through interlock switches. A further source of hazards is the active medium, especially in the case of several gas lasers, since gases can leave the laser and then cause harm, if they are poisonous or aggressive. A typical example is the excimer laser mentioned above, since the latter uses halogens as chlorine and others, that can etch eyes, lungs and also skin. So severe safety precautions must be taken to prevent a deliberation of halogens, as for instance safe gas supply systems, fixed metallic tubes instead of flexible hoses and of course storage of gas bottles in safe and locked shelters and the immediate availability of gas masks in the case of an accident. Finally, the third main assembly of a laser, namely the optical resonator with its at least two mirrors, is only a source of hazards as far as it concerns the shape and the direction of the beam, for which the resonator is responsible. Hazards connected to these beam properties are treated below. In the case of CO_2 lasers the output mirror is usually made from Zinc Selenide. If the latter is destroyed for any reason, Selenium is set free and very poisonous asking for evacuation of people and use of gas masks.

6.2.2 *Hazards of high power laser beams*

Due to their small cross-section, laser beams carry highly concentrated light energy, that can hurt the human eye, especially by a destructive overload of the photo receptors or even by burning and can also do harm to the human skin, at least as a sun burn or even by more severe burning. These threats of laser radiation are enormously enhanced, if the ultraviolet radiation, as for instance generated by excimer lasers, is considered. In the case of the human eye the dangers of laser radiation are multiplied due to the focusing properties of the human lenses and the excellent focusability of laser light (between 400 and 1,400 nm), leading to strongly magnified intensities acting on the photoreceptors. All these dangers do not only appear, if the direct laser beam hits the human body, but also if radiation is reflected, which can happen if bending mirrors on the way of the laser beam between the source and the workpiece are misaligned or if the laser beam hits highly reflecting material, that is not safely fixed, or finally if during a process reflecting surfaces as melt or plasmas are formed, the latter especially in deep penetration welding as mentioned above. Besides this dangerous reflection of the direct beam also the emission of secondary radiation can take place, even with lower wavelength, that is more dangerous, if it is in the ultraviolet, as it has also been treated before. Besides a reflection of the direct beam or secondary emissions also diffuse reflection, where the reflected wave is no longer a sharply collimated beam, but shows a broad spatial distribution, can be harmful, if the initial laser beam is of high power.

To judge on the danger potential of a certain given laser, lasers were for a long time divided into four main classes, where in class 1, covering laser powers in the order of microwatts, even looking into the direct beam is secure. In the case of class 2 lasers the beam power is increased, but they must be in the visible range, since in this case closing of the eye due to the irradiation limits the time of exposure. Finally class 4 lasers are so strong, that even the diffuse reflection, as mentioned above, can do harm to the eye. Nowadays, six classes are used (see Tab. 6.1). To be protected against the hazards of the laser beam, one must first find out the class of the laser, that must be indicated on the latter. If a class higher than 2 is stated, safety goggles must be used. These are small narrow band filters which match to the small band width of laser light, but make the use of different goggles for different lasers necessary, as for instance excimer lasers, Nd:YAG lasers or CO_2 lasers. Of course safety goggles must clearly indicate their wavelength of protection. Specifically dangerous is the situation where deep penetration welding with carbon dioxide lasers is carried out, since in this case strong ultraviolet radiation due to the formation of a plasma as mentioned above is emitted and so the safety goggles must protect against ultraviolet and against infrared radiation. Besides this direct protection of the eye, the whole laser system or at least those parts, that are reached by the laser beam, must be contained in housings, that prevent laser radiation to escape or people to get into contact with the beam.

Usually only the near surrounding of the processing zone, where the laser carries out its work, is enclosed, whereas the laser is standing separately, the beam being guided to the work station through metallic tubes or fibers. As in the case of protection against electrical hazards, these tubes and all other parts of the housing must be protected against removal by interlock switches. Of course the diameter of the tubes guiding the laser beam to the workstation must have at least twice the nominal beam diameter due to the intensity distribution of the beam reaching into infinity as mentioned above. Otherwise the beam would lose energy and the tubes would heat up dangerously. Concerning the housings of the laser processing system, doors that allow operating personnel to load and

Table 6.1   Laser Classes (Rockwell USA)

Laser class	Description	Application
1	Completely safe when viewed with the bare eye or magnifying optics	
1M	Completely safe when viewed with the bare eye but not with magnifying optics	
2	Only visible light (400–700 nm) and max.1mW CW or emiss.time 0.25s	Laser pointers
2M	Wide or diverging laser beams that limits the amount of light crossing pupil to limts of class 2	
3R	Max.5 mW for visible CW lasers, restricted beam viewing, low risk of injury	
3B	Big risk at direct exposure of eyes, asking for safety goggles, no risk at diffuse reflection	DVD writers
4	Highest risk of injuries of skin and eye, risk of the ignition of burning, even for diffuse reflection	Industrial lasers

unload the workpiece and to adjust the system must also be protected by interlock switches and systems, for instance weight sensitive carpets. These measures must ensure, that the laser can only be turned on, if nobody is inside the processing cell and if the doors are locked.

6.2.3 *Hazards of laser processes*

6.2.3.1 *Overview*

Laser processing can mainly be divided into three categories, the first one with essential material removal as in the case of cutting and ablation, the second one with material addition as in the case of welding or cladding and 3D printing and thirdly processes without a change of the workpiece mass in the case of hardening, bending or laser assisted forming. Each of these three categories is subject to specific hazards, that are in the first category associated with the removal of material. In the second category the hazards are associated to violent evaporation, that does not really lead to a loss of material, but gives rise to the formation of plasmas, that emit strong light, even in the ultraviolet, and may also lead to a reflection of the working beam. The third category seems to be much less dangerous, since neither material nor radiation is emitted with the exception of reflection and therefore little interaction with operating personnel takes place.

6.2.3.2 *Processing with material removal*

In this case a fast gas jet removes liquid material, that is resolidified after separation from the bulk of the melt and forms drops, if it originates from liquid material ejected at the bottom of the workpiece, as in the case of *laser cutting and ablation* or that resolidifies as a powder, if it is driven away from the surface of the liquid material due to friction with the gas jet. In the latter way also dust and aerosols can be emitted by the liquid material, that can be extremely dangerous as they may cause cancer in the lungs. Finally, if the liquid is hot enough to evaporate, metal vapor is formed. Solid material and vapor emitted during material removal can show a different chemical composition than the workpiece material, on the one hand due to the decomposition of molecules, what may result in the emission of poisonous products, and on the other hand it can be caused by reactions, for instance with oxygen used or contained in the gas jet, what can also deliberate dangerous materials, as for instance chromium oxide in the case of processing stainless steel. The material ejected from the molten layer, that is still liquid immediately after separation from the melt and shows a temperature above the melting point, can also do harm to the personnel by burning, if the ejected material hits the skin or even more dangerously the eye before resolidification and cooling down. This phenomenon of the ejection of liquid material of high temperature can be observed in laser cutting, where on the bottom side of the workpiece a so-called "spark shower," that consists of molten droplets with high temperature, appears. In laser ablation or laser suction (a process with melt removal by exhaustion), on the contrary, the spark shower is generated at the upper side of the workpiece and is directed, e.g., in parallel to the surface (see Fig. 6.2).

Fig. 6.1   Spark shower during laser suction cutting with the CO_2 laser (Institute of forming and laser technology, TUVienna, 1998).

Source: Authors.

In cutting and ablation of certain materials, that do not melt, but directly evaporate (sublimate), the pressure of the leaving vapor stream acts on the solid surface and induces stress that leads to the separation of fine particles from the surface and thus dust is generated. If evaporation takes place via the intermediate step of melting, due to the recoil pressure also small liquid droplets are splashed out and after cooling and resolidification also form dust and similar particles.

In ablative laser processing with material removal due to various mechanisms, solidified particles of either size from larger grain sized particles to very fines called "aerosols," and also vapor and gases, that might have a different chemical composition than the workpiece material leading to poisonous emissions are deliberated and form main perils of this category of processes.

Precautions that prevent particles emitted during material processing from doing any harm, comprise housing in of the workstation to prevent products to escape from the workstation and exhaustion of the latter by appropriate filters, that are matched to the nature of the various emission particles, powders, dust, aerosols and also vapor. Moreover, since material is ejected as initially liquid and hot droplets, that cool down and form solid drops, appropriate means must be used to collect these particles and allow their removal. Therefore specific traps do not only collect resolidi-fied droplets, but also catch that portion of the laser beam that is not absorbed by the workpiece and leaves the latter, e.g., at its bottom. So in cutting usually a metallic box is situated below the workpiece and catches there solidified droplets ejected at the bottom of the workpiece, thus pre-venting them to burn anything. Under specific circumstances this box can also be connected to an exhaust system that carries away dangerous parti-cle emissions, for instance in the case of laser cutting by suction through the cut kerf.

6.2.4 *Hazards associated to processing with material addition*

Material addition can be accomplished either by welding a first work piece to a second or by building up layers of initially powderized material on a workpiece as in the case of cladding, where thin layers finally cover the surface of the workpiece, or by building up three-dimensional struc-tures on a substrate in 3D printing.

In the first case of laser welding, especially in deep penetration weld-ing, the two work pieces fixed together in a butt geometry must be heated up to the evaporation point by the laser beam at the separation line in order to form a thin channel, extending throughout the full depth of the work-piece and thus allowing the laser beam to penetrate into the depth of the workpiece. Due to this mechanism of deep penetration welding the latter vapor channel blows out metal vapor especially at the upper side of the workpiece, thus giving rise to a vapor cloud above this channel. The latter can now absorb laser radiation, as mentioned before, leading to ionization and to the formation of plasma with a temperature well above 5.000°C. Although the formation of this plasma is desired, since it leads to a better coupling of laser radiation to the workpiece by *abnormal absorption*, it is associated with the dangers of strong UV emission, quite harmful to oper-ating personnel (Fig. 6.2).

This secondary emission is typical for all manufacturing processes, especially cutting and welding, where hot plasmas are formed and used and therefore the hazards associated with ultraviolet radiation are well known. The protection against this secondary radiation can be done by

Fig. 6.2   Plasma formation during laser beam welding (TU Wien, 1995).
Source: Authors.

housings, as mentioned several times before, and of course also most essentially by safety goggles, that have also been treated above.

Moreover, deep penetration laser welding is also dangerous, since the high recoil pressure generated by violent evaporation acts on molten surfaces, that are always present in the case of welding by definition, and therefore usually very hot and molten material splashes around, damaging on the one hand especially the focusing optics but on the other hand can also be harmful to operating personnel or cause a fire, if combustible materials are near to the welding system. Again housing of the workstation is the most important prerequisite for a safe operation. Concerning the destruction of optical elements by splashing melt, fast gas flows in the processing head, catching molten droplets and carrying them away safely, have been used successfully. In the case of laser cladding and workpiece generation from powderized material, a certain amount of powder is always lost into the environment of the workpiece and can do harm to the lungs of operating persons. Therefore again housing of the workstation and a connection of the latter to an appropriate exhaust system are also necessary for this kind of manufacturing processes.

Concluding, the most important safety precautions for material processing with lasers comprise mantling of the processing system, thus protecting the operating personnel on the one hand against the emission of dangerous particles, aerosols, dust and vapor and on the other hand against radiation, originating from the working laser beam either directly or reflected and also emitted by hot work piece material as secondary emission with eventually much lower and thus more dangerous wavelength. So with housings made of an appropriate material, that cannot be penetrated by emitted material and transmits neither laser radiation nor secondary emission, but is preferably at least in some parts transparent for visible light, thus allowing the operation personnel to observe the workstation, a safe operation of laser processing can be ensured, whereas also a well designed system of interlock switches gives additional security. So far all safety precautions mentioned before have only been designed to protect the operating personnel. Nevertheless it is also important to avoid a pollution of the environment by dust, aerosols or poisonous gases. Thus it is not sufficient to exhaust the work station as mentioned above, but appropriate filters must be used that catch emitted matter in various stages. They have to match to the size and nature of the dangerous emissions in order to prevent them from being deliberated into the environment. So filter techniques are also of major importance for a safe operation of laser material processing systems.

Chapter 7

Introductory Remarks
on Competing Technologies

Production technologies must be able to remove material as in cutting or milling, to add material as in welding or printing and to perform changes of workpiece geometry or properties without mass change as forming or hardening. All these tasks can be met with lasers, where with mechanical forces not all tasks can be performed as welding and printing and electrical processes can not be used for forming. Thus lasers cover practically all categories of production and mechanical, thermal and electrical technologies are less universal and restricted to distinct categories, but have an overwhelming importance and will therefore be treated below. Technologies as of optical, acoustic and chemical nature are restricted to a narrow range of applications and will not be treated here.

In order to provide a thorough understanding of mechanical technologies the next chapter containes a tutorial dealing with the effect of forces acting on solids. They can cause permanent, *plastic deformations* as they are necessary for forming and they can cause the *chipwise removal of material* as necessary for machining.

A second tutorial is then devoted to thermal technologies, that means the modification of workpiece properties, especially steel, by heating and cooling as they are crucial for hardening and related processes.

The third tutorial is then devoted to electrical phenomena that lead to heating and evapoation as they are necessary for erosion processes and welding.

These tutorials help to understand the various manufacturing processes treated in the following chapters, again divided in three groups with or without material removal or processes where the workpiece mass remains unchanged. All these processes will finally be compared to laser technology especially in view of speed, quality, safety of health, preservation of the environment, user friendliness, material and energy consumption, tool wear and of costs.

Chapter 8

Mechanical Technology (Tutorial)

8.1 Plasticity[1]

8.1.1 *Basic remarks*

Flawless single crystals with completely undisturbed crystal lattice are difficult to deform, because the bonds between the individual lattice atoms have to be broken in order to achieve a more far-reaching deformation than in the area of elasticity. This requires very large forces. In such a completely flawless lattice, in the absence of external forces acting on the material, there are no tensions, since all the lattice atoms are in the places where the attracting and repulsive effects of the individual atoms are canceled, which is true at least if one does not take into account the thermal movement of the atoms.

If the lattice of a crystal now shows deviations from the regular arrangement of the atoms, as in the case of an intermediate lattice half-plane (Fig. 8.1), this leads to the fact that not all atoms sit in the places where no force is exerted on them. Therefore, at least on a part of the atoms act forces, so that in the crystal tensions arise even in the absence of external forces. If forces initiated from the outside are now superimposed to these internal stresses in such a way that both reinforce each other, a tearing of individual lattice bonds can take place even at less forces acting on the workpiece than in the case of an undisturbed lattice,

---

[1]This chapter is based on a condensed version of chapter 4.1 of D. Schuöcker, Spanlose Fertigung, p.100–105, Oldenbourg, München 2004.

external shear stress

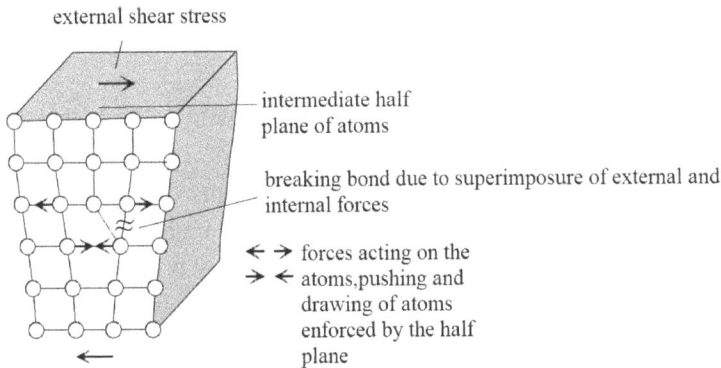

intermediate half
plane of atoms

breaking bond due to superimposure of external and
internal forces

← → forces acting on the
→ ← atoms,pushing and
drawing of atoms
enforced by the half
plane

←

Fig. 8.1 Intermediate lattice planes.

Source: D. Schuöcker, Spanlose Fertigung, Fig. 4.1.7 Oldenbourg, München 2004.

which facilitates a deformation. Thus, only the appearance of lattice defects in crystals opens up the possibility of reshaping these materials. Crystals in undisturbed lattice can hardly be reshaped.

However, not all possible types of lattice defects contribute to a lighter deformability, whereby initially point-shaped errors of the crystal lattice such as intermediate lattice places or substitution faults do not play a significant role. Only surface lattice defects, such as intermediate lattice planes and displacements, are important. This is because they can migrate from one side of the crystal to the opposite side when moving through the crystal as a result of external forces, whereby the workpiece on the first side loses material and gains material on the other side, so that a deformation occurs overall.

Subsequently, a kind of lattice faults, namely intermediate lattice planes, is to be discussed, whereby it is first explained why such a lattice error can move through the crystal under the influence of external forces, along which plane and in which direction this movement runs, how such crystal defects multiply by mechanical deformation or by increasing the temperature, how they interact with each other and how they behave at the grain boundaries.

Thus, all basic phenomena of forming technology, such as cold and hot deformation, hardening during deformation and recrystallization and reversal of the change in the structure associated with the deformation can be explained.

8.1.2 *Intermediate lattice planes and their mobility in the crystal*

Intermediate lattice planes are atomic planes, only half of which are occupied by atoms (Fig. 8.1). In the vicinity of the end or edge of the intermediate planes, forces are now exerted on the atoms of the adjacent grid planes in such a way that the adjacent and complete lattice planes approach each other with rising distance from the end of the intermediate lattice plane and finally take their distance in the undisturbed crystal again at a greater distance from the edge of the lattice disturbance. Thus, in particular, the atoms to the right and left of the intermediate plane and above their end are under pressure, while the atoms are under tensile tension just below the edge of the intermediate grid plane.

If now, as in Fig. 8.1 exemplified, on the crystal acts a pair of sheer forces, the pressure on the atoms, which are located to the right above the edge of the intermediate lattice plane, is intensified. Likewise, the pull is amplified on the atoms located to the right below the end of the intermediate lattice plane, so that overall on the two considered atomic series just above and just below the end of the intermediate lattice plane and right of it acts a strong, loosening force, whereby this lattice bond can be broken even at a shear load, which lies far below the force necessary for a deformation of an undisturbed single crystal. Thus, the atoms at the free end of the intermediate lattice plane can now enter into a bond with the atoms to the right below the end of the intermediate plane by the above-described tearing of the atomic pairs, so that the previous half-lattice plane is completed, and the previously right-hand and complete plane becomes a new intermediate plane, whereby the intermediate plane is moved to the right by an atomic distance. Thus, an intermediate lattice plane can be moved through the crystal when applying a shearing load, the strength of which is far lower than it would be necessary for tearing the lattice bonds in an undisturbed crystal.

The plane that crosses the end of the lattice plane is called the *sliding plane*. If an intermediate lattice plane along a sliding plane moves from the left edge of the workpiece to the right edge in this way, material is mined from the left edge of the workpiece and built up at the right edge, which means that the workpiece has been deformed along the sliding plane, whereby the necessary forces are far less than those that would be necessary to break an undisturbed, errorless crystal by breaking up. The shear stress per unit area that causes the beginning of permanent, *plastic deformation*, is called the *yield strength*.

The latter is not represented by one value for steel, but is represented by a narrow range between the upper and lower limits (various steels: lower limit 200–450 N/mm^2).

For the actual deformability of a crystal afflicted with lattice defects, it is now crucial how many intermediate lattice planes are present per volume unit, how large their mobility is in the crystal, how they interact with each other and how they behave at the grain boundaries. First of all, the question of the interaction of the lattice faults among themselves is to be discussed.

8.1.3 *Interaction of the intermediate grid planes with each other*

If one looks at two adjacent intermediate grid planes, which are coincident and are in the same workpiece part (Fig. 8.2), it is shown that they exert a repulsive effect on each other. This is because they exert pressure forces on the atoms of the complete planes between them from both sides, which causes reaction forces to act on the atoms close to the

intermediate lattice plane

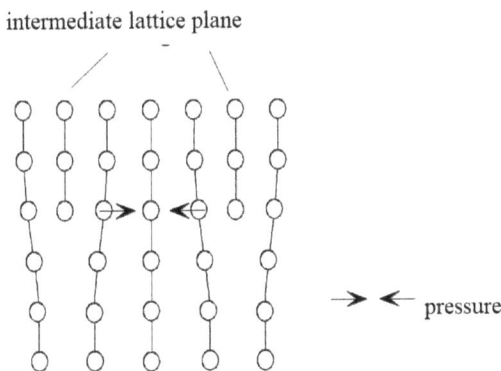

Fig. 8.2 Repelling action between two intermediate lattice planes.

Source: D. Schuöcker, Spanlose Fertigung, Fig. 4.1.10 Oldenbourg, München 2004.

two intermediate grid planes considered here, which seek to push the two levels apart.

Similarly, it can be explained that two intermediate grid planes, which are the opposite of each other, attract each other. Due to this attractive interaction, two intermediate grid planes can recombine and form an undisturbed atomic plane.

8.1.4 *Agility of the intermediate lattice planes*

The repulsive effect of two coincident interlattice planes has the effect that the mobility of the intermediate lattice planes in the crystal is limited by a large number of such crystal defects, because the movement of one half plane hinders that of the other. The mobility of the intermediate lattice planes thus decreases considerably with their number per volume unit. On the other hand, the mobility of the intermediate grid planes is greatly facilitated by increased temperature, because then as a result of the stronger thermal movement of the atoms the lattice bonds are loosened and thus the tearing of the bonds, which is necessary for the movement of the intermediate lattice planes, is facilitated. For this reason, the deformability of most materials at higher temperatures is also significantly improved.

8.1.5 *Behaviour of the intermediate grid planes at the grain boundary*

Practically used metals are in the polycrystalline state, i.e. individual grains with a structure similar to the single crystal are separated from each other by grain boundaries, wherein these crystallites have different orientations of the sliding planes (Fig. 8.3). If an intermediate lattice plane migrating through a grain reaches a grain boundary, its further propagation is hindered by the fact that the crystallite present behind the grain boundary has different orientations of the sliding planes. In case of slight deviations, the intermediate grid plane may be able to overcome the grain boundary and move on in the new crystallite. However, if the difference in orientation of the two crystallites is too great, a further migration of the intermediate grid plane is severely obstructed to impossible.

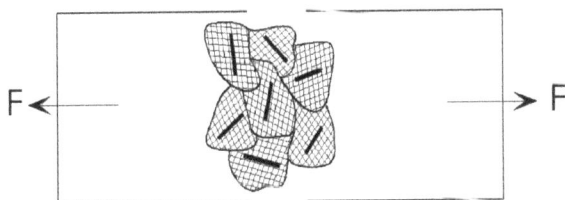

Fig. 8.3   Grains subject to tensile stress.

Source: D. Schuöcker, Spanlose Fertigung, Fig. 4.1.11 Oldenbourg, München 2004.

The more grain boundaries the material has, the finer the structure, the lower the mobility of the intermediate grid planes and therefore also the deformability of the material. In the case of coarse grains, however, the material is easier to deform.

8.1.6 *Work hardening and recrystallization*

Since the mechanism of deformation of materials by external forces has been explained, the change in the properties of the material in the course of deformation is now to be discussed in more detail:

Very often, a workpiece is not subject to shearing forces, but by either tensile or compressive forces. In the individual crystallites, however, the sliding planes are generally not parallel to the acting tensile or compressive forces (Fig. 8.3), but inclined to these, whereby then both induce shear forces in the sliding planes, which, however, have different size in the different grains due to the different orientation of the sliding planes. This makes some grains stronger and others weaker deformed, which then leads to a deformation of the entire workpiece.

By the above mentioned deformations, the initially isotropic grains are either stretched in the direction of the tensile forces or compressed in the direction of the compressive forces (Fig. 8.4), whereby a strong anisotropy is generated in the structure. In detail, the number of atoms per unit volume and their distance cannot change during the deformation of the workpiece and the volume of the grains must therefore remain constant.

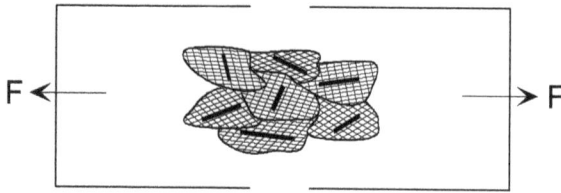

Fig. 8.4   Deformation of crystallites due to tensile stress.
Source: D. Schuöcker, Spanlose Fertigung Fig. 4.1.12 Oldenbourg, München 2004.

Thus, the grains are constricted perpendicular to the pull direction or widened perpendicular to the pressure direction. The movement of the lattice faults along the sliding planes is more strongly impeded by the grain boundaries, since the lattice errors reach the next grain boundary already after a shorter movement distance. However, the deformed grains make further deformation more difficult, i.e. the previous deformations reduce the deformability and increase the hardness and strength. This effect is called *work hardening*. If the workpiece is now heated to a temperature of a few 100°C, dissolution takes place of the grains deformed by the deformation of the workpiece and new small grains begin to form.

This process occurs with most metals at a temperature of one third of the melting point and takes a few seconds. If the workpiece is kept at the above temperature for a few minutes or even longer, the initially very fine new crystal structure becomes coarser and coarser, which means that the strength also decreases again and finally assumes the value that it took before cold solidification, a phenomenon called *recrystallization*.

8.1.7 *Application of plasticity*

Plasticity allows semi-finished products such as sheets or wires to be brought into a specific shape without mass modification. This forming can be carried out by various forming tools, such as matrices and stamps, which exert on the workpiece the necessary forces for the flow. Often this transformation is only possible in several steps, because after each step work hardening results during the forming, which hinders further forming.

It can be reduced by annealing in a furnace, which greatly increases the production time.

The use of materials here goes up to almost 100%. One example is deep drawing, which is widely used in the manufacture of automotive body parts.

8.1.8 *Equipment for forming*

Figure 8.5 shows an overall view and 8.6 the construction details of a small deep drawing press with a press force of about 60 kN. The machine consists of the press table, on which four guide columns are mounted. At the other end of the guide columns, the transverse head is attached that carries the hydraulic main cylinder. The main cylinder shifts the pusher of

Fig. 8.5 Hydraulic forming press.

Source: Authors.

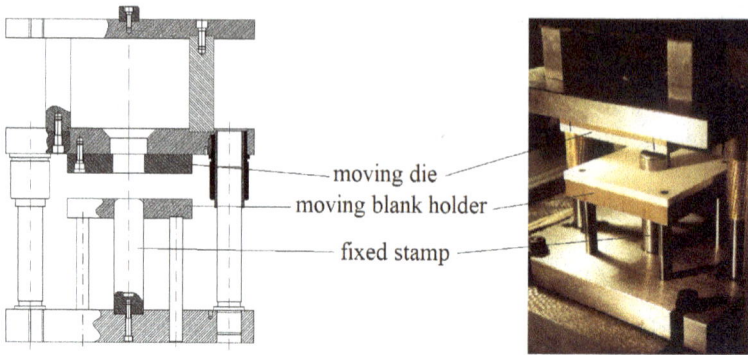

Fig. 8.6   Deep drawing press inside view.

Source: Authors.

the press on the guide columns. The maximum operating pressure of this press is 350 bar. The pusher carries the drawing die. The stamp (Fig. 8.6) is mounted on the press table. Under the press table there is another hydraulic cylinder, which serves to move the blank holder.

Such a deep drawing press is used in particular as a so-called "tool trial press". It is used in tool production to test the tool as well as for the production of small series.

For series production, in which many strokes follow each other quickly, instead of a hydraulic drive, usually a continuously rotating eccentric is used, which turns the rotation into an up and down movement of the stamp.

The tools for deep drawing are made of tool steel and must be hardened. Their production is usually very expensive.

8.2 Fatigue and Break

8.2.1 *Tensile and shear strength*

If a workpiece is applied with an ever-increasing force after vigorous plastic elongation, its cross-section continues to tighten, as its mass must remain equal. At the same time, it solidifies more and more until an increase in strength cease hardly any further elongation. However, small defects on its surface, such as notches, begin to grow due to the high tensile stress. This leads to an instability in which the workpiece cross-section is reduced by the the cross-section of the notch, which increases the tensile tension, which in turn strengthens the lacing; a mutually reinforcing interaction which now proceeds without further intervention from the outside and leads to complete lacing and thus to breakage. The elongation at breakage, the ductility is about 20% for the various structural steels and the necessary tensile stress, the tensile strength about 300–800 N/mm^2. Breakage, of course, also occurs in shearing load (shear strength). In the case of steel and shear stress, rupture takes place at approximately 80% of the tensile strength.

8.2.2 *Application to machining*

The fracture occurring upon reaching the tensile or in particular the shear strength allows workpieces to assume the desired shape by material removal (machining). This removal can be carried out by tools which are either shape-determining (e.g. drills) or which determine the desired geometry by the relative movement with respect to the workpiece (e.g. rotary chisel), whereby material is removed by combined effect of pressure and shear in the form of chips.

Chapter 9

Thermal Technology (Tutorial)

9.1 The Iron–Carbon System Dependent on Temperature[1]

Pure iron is weak and cannot be used for technical applications. Nevertheless if even small amounts of carbon are added, steel with high strength is obtained. The latter is very sensitive to only small changes of the carbon content and alters its microstructure and properties if the temperature changes. The iron–carbon diagram now shows for equilibrium state the various phases of the binary Fe–C system for given carbon content and for a certain temperature. To be able to read the latter diagram the various phases are explained with their properties in the following (Fig. 9.1):

9.1.1 *Ferrite*

Two phases are δ-iron at temperatures above 1,392°C and α-iron below 911°C.

Ferrite has a body centered cubic structure up to a temperature of 723°C with iron atoms in the corner and one in the middle of a cube.

It has low heat expansion, is easy to deform (ductility 50%), machining is characterized by low forces necessary, but leads to the formation of

---

[1] Sections 9.1 and 9.2 form an essence of relevant articles from Wikipedia, the free encyclopedia, as on the Iron–Carbon diagram and the Time Temperature Transformation diagram.

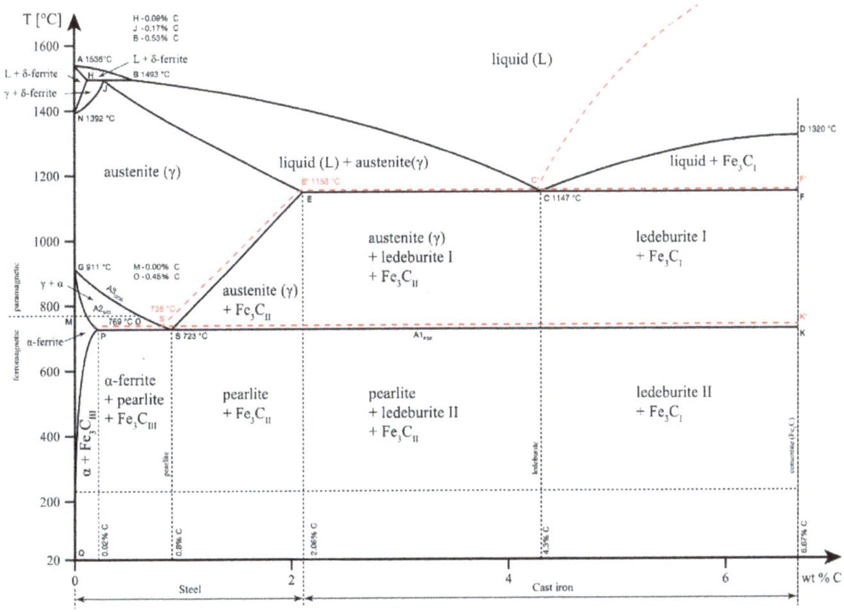

Fig. 9.1 Fe C diagram (By AG Caesar — Own work Läpple, Volker — Wärmebehandlung des Stahls Grundlagen, Verfahren und Werkstoffe 8. Auflage, Seite 55ff. Weißbach, Wolfgang — Werkstofkunde Strukturen, Eigenschaften, Prüfung 17. Auflage, Seite 76ff. http://www.chemie.de/lexikon/Eisen-Kohlenstoff-Diagramm.html#Darstellung_der_ Phasen_im_Eisen-Kohlenstoff-Diagramm, CC BY-SA 4.0, https://commons.wikimedia. org/w/index.php?curid=76922696)

burrs and yields low surface quality. Also the formation of built-up cutting edges (see Chapter 10) is likely.

Since it can only dissolve very few (0.02%) carbon, the ferrite grains pervade laminae of **cementite**, that is iron carbide Fe_3C, a hard and brittle compound. The latter mixed phase leads to a mother of pearl like appearance and is called **pearlite**.

If the carbon content of steel is 0.8%, above the transformation temperature of 723°C according to the Fe–C diagram the lattice changes after a short holding time (arrest) to **austenite**, also γ iron, with a face centered cubic structure with iron atoms at the eight corners and also one in each face center (six) of a cube. During this transformation, cementite breaks into iron and carbon, the latter being dissolved by the austenite, that can dissolve much more carbon than ferrite. The carbon atoms are then arranged interstitially between the regular sites of the iron atoms and push them away from these equilibrium sites where they do not experience forces (thermal movement neglected) what leads to forces acting on them quite similar to a spring extended from its equilibrium length. This stress acting on the lattice caused by the carbon atoms leads to enhanced hardness.

For a carbon content of 0.8% the two phases of pearlite under 723°C change directly to one phase of austenite above 723°C. Since at that point carbon is 100% dissolved in austenite and thus only one phase remains quite similar to materials that have a distinct melting point instead of a dough like intermediate region (*eutectic* material) and is thus called *eutectoid* steel. By the way, austenite can also be found at room temperature, especially in the case of stainless steels.

The latter phase is non-magnetic, shows high ductility, high durability and excellent corrosion resistance. These steels are considered as the most difficult-to-cut materials as compared to the other alloy steels due to their high work hardening, low heat conductivity and high built up edge formation.

Heat treatment of austenitic steels by quenching their temperature quickly leads to the formation of a very hard phase, **martensite**, that will be treated below.

Pearlite can be the sole constituent of eutectoid steel (0.8% C) and can be used for machine and hand tools.

The tensile strength is much higher than for construction steel (800 N/mm^2) and the ductility 48%.

Thus, it is well suited for machining as well as forming. Thin wires drawn from hypereutectic pearlite can be made extremely strong, the tensile strength reaches 6,000 N/mm^2. In machining the large hardness causes higher forces and enhanced wear and the formation of built-up cutting edges, although the quality is good.

Hypo and hypereutectic steels can be made eutectic with sole pearlite composition by special heat treatment prior to machining.

Martensite is formed by quick cooling (quenching) of the austenite to prevent carbon atoms from diffusion out of the crystal structure to form cementite (Fe_3C). As a result of the quenching, the face-centered cubic austenite transforms to a highly strained body-centered tetragonal form called martensite that is supersaturated with carbon. The shear deformations that result produce a large number of dislocations, which is a primary strengthening mechanism of steels. The hardness of martensite can achieve 700 Brinell (pearlite 400 Brinell).

For a eutectoid carbon steel sheet, if the quenching starts at 750°C and ends at 450°C during 0.7 s (a rate of 430°C/s) no pearlite will form, and the steel will be nearly completely martensitic with small amounts of retained austenite. Martensite show practically no ductility and is thus very brittle.

Cast iron is a group of iron-carbon alloys with carbon content more than 2%. Its wide application for casting is determined by its relatively low melting temperature and its low viscosity melt. Nevertheless it is quite brittle and the formation of cracks is likely, where *white cast iron* has carbide impurities which allow cracks to pass straight through, *grey cast iron* has graphite flakes which deflect a passing crack and initiate many new cracks as the material breaks, and ductile cast iron has graphite *spheres* which stop the crack. The latter material is well suited for machining. Due to its brittleness forging is not possible.

A cast iron with a carbon content of 4.3% is referred to as eutectic cast iron. The main constituent is **Ledeburite** with nearly equal content of cementite and pearlite (Ledeburite II) or austenite (Ledeburite I). Details can be seen from the above Fe–C phase diagram.

9.2 Time Temperature Transformations

Phase transitions as from pearlite to austenite or from austenite to martensite, the latter most important for heat treatment, as hardening, take place at a certain temperature after immediate cooling. The latter is kept constant (isothermal) and transformation starts after a certain time, grows and reaches a maximum.

The latter process is described by a *Time Temperature Transformation* (TTT) diagram for a distinct material as above for the example of an eutectoid steel with 0.8% carbon (Fig. 9.2).

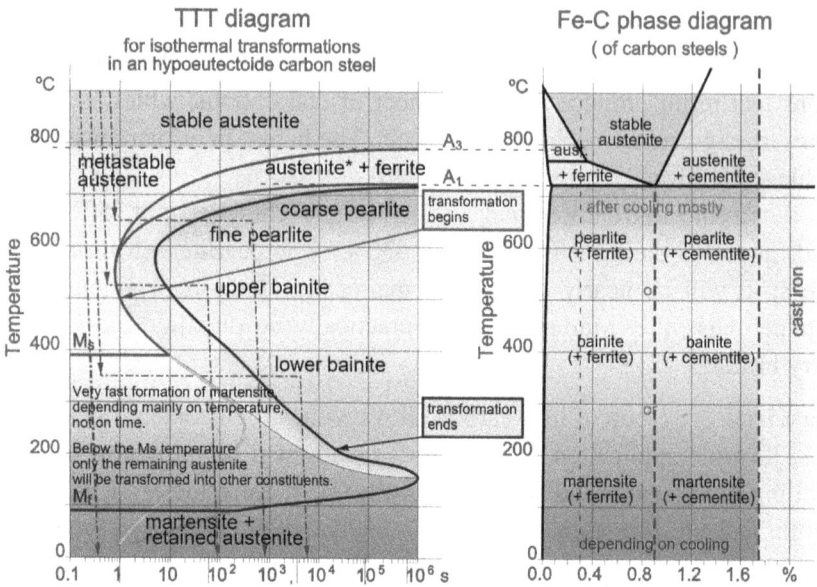

Fig. 9.2 TTT diagram for hypoeutectoid carbon steel (Bainite is steel with needles of cementite embedded in ferrite, M_s indicates start of transition, M_f finished transition to martensite).

Peli Oleaga-Own work, CC BY-SA 4.0, https://commons.wikimedia.org/w/index.php?curid=95989750

It shows at the vertical axis the temperature and at the horizontal axis the time after immediately establishing a certain temperature and then keep it constant. For a given temperature the transformation from austenite to pearlite to starts at the leftmost curve and reaches a maximum percentage at the rightmost curve. Close to the transformation point the curves approach the latter asymptotically since at that temperature of cause no transformation can start.

The martensite transition occurs immediately and the percentage of martensite depends only on the temperature and remains then constant independent of time and is shown by horizontal lines, the uppermost for starting temperature and finally for the finishing temperature a maximum percentage that is smaller than 100%.

Chapter 10

Electrical Technology (Tutorial)

10.1 Plasma

10.1.1 *Ignition of a plasma*

Gases are usually very good insulators and can conduct virtually no electricity. However, if two metallic electrodes are located in a gas at a certain distance (see Fig. 10.1) and connected with the positive and negative poles of a voltage source, an electric field builds up between the electrodes. If the field strength is large enough, the few electrons generated by high-altitude radiation, which are accelerated towards the positive electrode, can gain so much energy between two collisions with atoms that they supply the ionization energy to the spelled atom at the next collision, thereby ripping off an electron from the atom. The bumping electron is retained, so that after the ionizing shock two electrons move towards the positive electrode. The two electrons are then accelerated again and perform further ionizations of atoms, whereby finally a constantly swelling *avalanche of electrons* moves towards the positive anode (see Fig. 10.1). By the way, after each ionization positively charged atoms — ions — remain, which then move towards the negative electrode, the cathode. After a short time, a part of the gas atoms is then split into negative electrons and positive ions — charge carriers that can transport electricity — and a weak current flow between the electrodes occurs, which is called the *ignition* of a gas discharge. The word *gas discharge* is due to the fact that such ignitions were originally carried out by means of capacitors charged

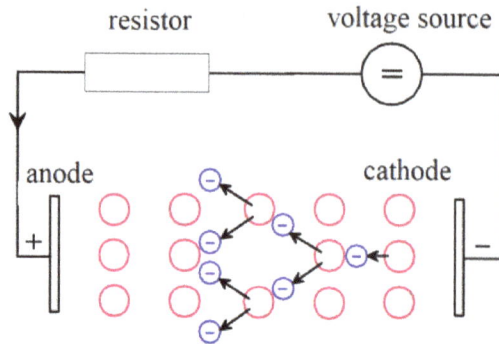

Fig. 10.1   Electron avalanche.

Source: D. Schuöcker, Spanlose Fertigung Fig. 1.4.1, Oldenbourg, München 2004.

to high voltages. The ionized gas with free electrons and positive ions is called *plasma*.

The ignition voltage increases with the electrode distance and the pressure of the gas, because field strength times distance between two atoms (force times way) indicates the energy gain of the electron, which must be equal to the ionization energy. Field strength times electrode distance gives than the ignition voltage in the order of 1,000 V for air at an electrode distance of 1 mm.

As shown in Fig. 10.1, in order to avoid excessive currents, which would lead to the destruction of the electrodes, a resistor is switched between the voltage source and the electrodes, so that only a small current can flow after the ignition and formation of the plasma.

If this resistance is reduced after ignition, the voltage drop at the latter decreases and at constant voltage of the power supply the voltage increases at the electrodes, whereby also the field strength between the electrodes increases and the electrons flow faster to the anode and the ions faster towards the cathode, which means that the current flow increases.

If the positive ions reach the negative cathode at a higher speed and therefore greater kinetic energy, they thus supply it with energy, so that it heats up and the electrons, which are fed by the current flow from the voltage source of the cathode, gain enough kinetic energy in the cathode to free themselves from the attraction of the positive atomic nuclei in the lattice of the cathode and thus leave the cathode. The current transfer from the cathode to the plasma thus takes place by energy supply to the electrons in the cathode. The electrons emitted at the cathode then move in the plasma towards the positive anode, but they lose their kinetic energy due to collisions with gas atoms and therefore have to be accelerated again by the electric field between the electrodes. In addition, the number of electrons moving from the cathode to the anode suffers some losses, as the negative electrons are captured by positive ions and recombine with them to neutral gas atoms. For this reason, additional charge carriers must be generated by shock ionization of neutral gas atoms either by electrons accelerated by the field strength or by gas atoms that move rapidly as a result of the current flow. The electron current then reaches the anode, where the electrons can easily enter the metal surface, so that the electrons can then flow from the anode to the voltage source.

As far as the emission of electrons from the cathode and the type of shock ionization inside the plasma are concerned, a distinction must be made here according to the gas pressure.

10.1.2 *Influence of gas pressure*

If the **gas pressure is relatively small**, only a few gas atoms and therefore also positive ions are present per unit volume and the energy supply to the cathode by the latter is so low that there is no significant warming. In this case, the electron emission takes place in such a way that the attraction between the electrons in the metal and the positive ions approaching the cathode becomes so strong that it outweighs the attraction of the electrons by the positive atomic nuclei in the metal. Since a few positive ions are present per unit volume due to the small pressure and correspondingly also a few negative electrons, the current per unit area is relatively small, so that only a weak heating of the gas takes place between the electrodes. Thus, ionization by the collision of rapidly moving gas atoms is out of the question and the shock ionization takes place only by the free electrons accelerated by means of the electric field strength between the electrodes.

The shock ionization by fast electrons, which takes place at low gas pressure, is due to the fact that these are strongly accelerated when passing through the free path length — relatively large due to the small pressure — and rarely collide with the few gas atoms and lose energy in the process, so that they assume a high kinetic energy that is greater than the ionization energy and therefore allows them to ionize atoms. This high kinetic energy can be characterized by a high electron temperature, which can be many $1,000°K$ under the conditions of low pressure, while the gas between the electrodes reaches only a temperature of little above $100°$.

The situation is quite different with **relatively high gas pressures** in the atmosphere range.

Here there are a lot of gas atoms and thus also positive ions per unit volume and the cathode is exposed to an intense bombardment by the latter, so that it heats up strongly, whereby the electrons in the metal assume such a high kinetic energy that they can free themselves from the attraction by the positive atomic nuclei and thus leave the cathode. The cathode can certainly assume temperatures of a few thousand degrees.

The numerous gas atoms present at high gas pressure per volume unit and thus also numerous positive ions and negative electrons lead to a high current density, which leads to the generation of heat and thus to a strong temperature increase of the gas between the electrodes as a result of the "friction" of the charge carriers with the gas atoms. At high pressure, the gas can assume temperatures of up to a few $1,000°K$, whereby the electrons reach about the same temperature, since they are "thermalized" due to the very frequent collisions with gas atoms due to the high pressure and the small free path length, which means that temperature differences between the two types of particles are compensated.

10.1.3 *Glow discharge and arc*

At **low gas pressure** and only few gas atoms per volume unit, only a few collisions between electrons and gas atoms take place, as already mentioned, whereby these shocks not only involve the necessary ionization — as described above — but also the stimulation of energy levels. The excitation energy supplied by the impact of the electrons is then released again in the form of light. For this reason, the plasma shows a glow. Since there are only a few gas atoms per volume unit, this light emission is relatively weak, so one speaks of a *glow discharge*.

On the other hand, under the conditions of **high pressure**, there are a lot of gas atoms, which also have a very high kinetic energy as a result of the high temperature described above, with which many and violent collisions take place between the gas atoms. In these collisions, in addition to ionizations, stimulations of higher energy levels are now carried out again, which leads to the emission of light similar to the glow discharge. However, since the number of gas atoms involved is much higher here than at low pressure, an intense luminous phenomenon appears. Under the conditions of high pressure, as said, the current density in the plasma becomes very high. However, with a given resistance between the electrodes and the voltage source, the current flow via the gas discharge is limited, so that the cross-section of the current flow must be reduced. This results in a thin, column-like plasma at high pressure, which leads from the cathode to the anode. This plasma column resembles an electrical conductor connecting the electrodes, which generates a magnetic field, which in turn exerts a force (*Lorentz force*) [1] acting on the electrons flowing in the conductor, which moves the conductor towards the edge of the electrodes at a sufficiently high current and then, since its electrode foot points cannot move further, bends outwards in an arc, thus making the term *arc* or electrical *arc* comprehensible for the plasma at high pressure (Fig. 10.2).

The contraction of the current flow described above does not occur at low pressure, at which a glow appears, so that it fills the space between the current-carrying electrodes homogeneously (Fig. 10.3).

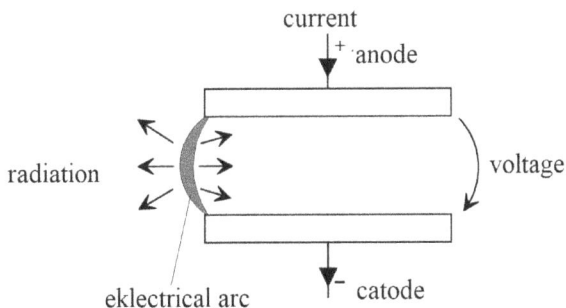

Fig. 10.2   Electrical arc.

Source: D. Schuöcker, Spanlose Fertigung, Fig. 1.4.2 Oldenbourg, München 2004.

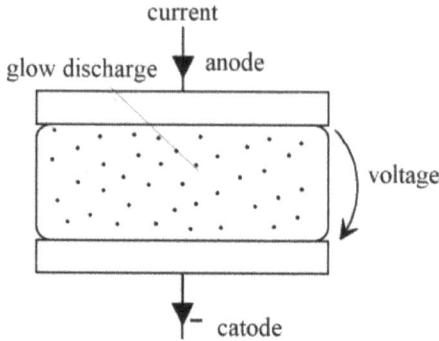

Fig. 10.3   Glow discharge.

Source: D. Schuöcker, Spanlose Fertigung Fig. 1.4.3, Oldenbourg, München 2004.

Due to the high gas temperature in the arc, a very strong radiation not only of visible light but also in the UV range takes place. The energy balance of the arc therefore includes the energy supplied by the current flow and the energy lost by radiation and energy supplied to the two electrodes.

Practical values represent a curve length of 5 mm, an arc radius of 1 mm and the minimum required temperatures of $T = 5,000°C$ for shock ionization. This results in a voltage of 21 V at a current of 500 A (radiated power 111 W, the anode supplied 2,800 W and the cathode power 7,500 W, all approximate values).

10.1.4 *Spark erosion*

The electrically conductive workpiece is arranged together with an electrode serving as a tool (see Fig. 10.4) in an insulating liquid, a non-conductive "dielectric." Between the tool electrode and the workpiece, the pulsating voltage generated by a power source is now applied, which is so high in the pulse peaks that a gas discharge is ignited, whereby first the liquid dielectric evaporates and the ignition takes place in the vapor. The power source provides such a high current that an arc discharge appears. This arc discharge heats the base points on the workpiece and the tool electrode so strongly that material evaporates there and from both electrodes material is removed in the form of a small crater. By appropriate combining the work piece material and the material of the tool electrode,

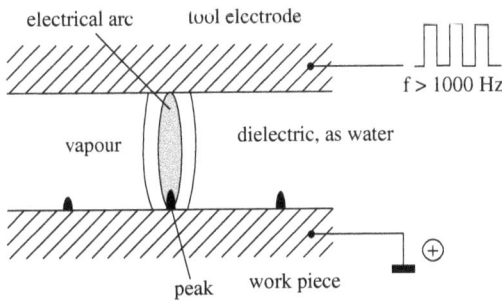

Fig. 10.4 Mechanism of spark erosion.

Source: D. Schuöcker, Spanlose Fertigung, Fig. 2.1.1 Oldenbourg, München 2004.

removal can practically only be achieved on the workpiece. The arc goes out after a short time of a few milliseconds at the end of the pulse, so that no continuously burning arc, but only a spark comes into place.

When the next voltage pulse arrives, the ignition of a spark takes place again at a point with preferred conditions. These would be hot spots on the electrodes, where electron emission is facilitated, or at a particularly small distance between the electrodes, where the electric field strength is especially high. Hot spots on the electrodes naturally come from the foot points of the last ignited spark, so that the sparks would ignite again and again in the same place and thus a removal would only take place at one point of the work piece. This is prevented by the cooling effect of the dielectric. This means that the next spark can only ignite at the point where the distance between the workpiece and the tool is still the smallest and where the workpiece has not yet been removed as strongly as in the surrounding area. This means that the spark ignites again and again at each pulse in a place where too little has been removed and a very uniform removal of the workpiece occurs. In order to achieve good cooling by the dielectric, it flows through the workpiece and the tool, which also allows it to remove the removal products. Since the dielectric is heated on the one hand and contaminated on the other hand, it is filtered and cooled in a closed circuit.

Reference

1. Wikipedia, the free enclopedia, "Lorentz force".

Chapter 11

Competitive Manufacturing Processes with Material Removal

11.1 Overview

Cutting of sheet metals is one of the most important manufacturing processes since every day products as cars or appliances mainly are made from sheets or contain sheets. Various options can be used: Mechanical as *shear cutting, nibbling* that means subsequent punching little holes finally resulting in a contour or *saw cutting*. Electrical cutting can be performed with *arcs*, that heat the material up to melting and also *sparks* that evaporate material and thus both yield material removal. *Water jets* can also be used to remove material due to abrasive effects.

Ablation, that means removal of workpiece material not only in thin sheets as in cutting but also in the depth of bulky workpieces, can also be performed with evaporation by spark erosion but is mainly done mechanically by machining, especially *drilling, turning* and *milling* and also *grinding*. Since turning is not really a competitor to ablation and drilling that can be regarded as an option of milling, this process is not treated here.

11.2 Cutting Processes

11.2.1 *Shear cutting*

Shear cutting (Fig. 11.1) is carried out with a stamp and a die with an opening slightly larger than the stamp to allow the flow of the work piece material into the die. The stamp moves the part to be cut, a piece of sheet metal, towards the die, thus drawing it a little bit into the die. In this situation a pair of shear force act on the workpiece just submerged in the die, one of the forces at the edge of the die and a second at the edge of the stamp. If the force on the stamp and in consequence the shear forces are strong enough, the work piece breaks between the edges mentioned above and a cut is obtained.

11.2.2 *Nibbling*

Arbitrary contours as with lasers cannot be cut in the above way but with small stamps little holes can be punched that can form an arbitrary contour if they sit along the latter touching each other, a process called **nibbling** (Fig. 11.2). Unfortunately the cut edge shows a saw tooth structure caused

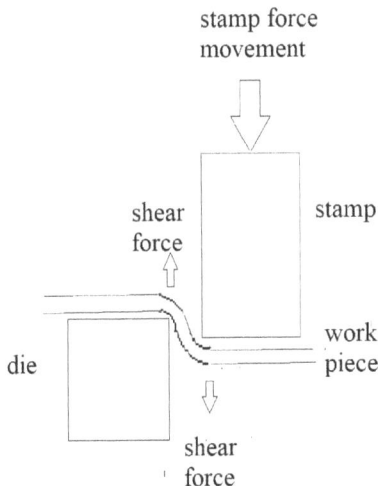

Fig. 11.1   Principal function of shear cutting.

Source: Authors.

Fig. 11.2   Schematic view of a workpiece during nibbling.

Source: Authors.

by the shape of the stamp. The quality of the cut can be improved, if stamps with a shape matched to the desired cut are used as, e.g., rectangular for straight lines or round for narrow curvature. A further disadvantage is that the process is very loud and generates a sound similar to a machine gun. Also strong forces act on the workpiece as well on the tool, thus enhancing wear and possibly damaging the processed part. Of course the tools are made from high strength tool steel and also hardened, processing speed can be up to 10 m/min dependent on the sheet thickness, that can be up to 10 mm.

Compared to laser cutting, nibbling is considerably slower but can cover much thicker workpieces. Material loss is bigger due to the wide cut kerf determined by the size of the tool stamp and it is less comfortable due to its noise. The quality of the cut edges is considerably lower, what makes usually post processing necessary. Moreover its performance depends on certain material properties, as hardness, but laser cutting depends on material properties, as heat conductivity. Nevertheless all these disadvantages may be compensated by a lower investment, that concerns only the energy source of the process and neither workpiece and tool movement nor CNC equipment.

11.2.3 *Saw cutting*

Mechanical cutting of metals can also be performed with metal saws that can be distinguished from wood cutting saws by the properties of the teeth, that are rather fine due to the much higher strength of metals but arranged to a high number per unit length.

The latter can be carried by a rotating round blade allowing only straight cuts but also on narrow bands with a forth and back movement

Fig. 11.3   Scroll saw for cutting arbitrary contours.

Dremel.jpg"></a>
By Scott Ehardt — <span class="int-own-work" lang="en">Own work</span>, Public Domain, <a href="https://commons.wikimedia.org/w/index.php?curid=1193182">Link</a></p>.

that enables cutting of arbitrary contours where the curvature can be stronger if the blade is narrower (scroll saw, Fig. 11.3). Since then the strength of the blade is reduced only thin or soft metal sheets as from aluminum or copper can be cut.

Compared to laser cutting of strong metals as steel, the latter can only be cut at much lower thickness and with lower speed. The quality of the cut edges can be similar to that obtained with lasers. The kerf width is somewhat larger and so is the material loss. The process is noisy and tool wear is important asking for sharpening the blades frequently. Again, all these disadvantages are compensated by reduced investment avoiding expensive laser equipment and using a much cheaper motor to drive the saw blade.

11.2.4 *Plasma cutting*

An arc can be used to cut metals, whereby the latter melts the workpiece at the momentary end of the kerf (see Fig. 11.4). The melt is then expelled at the bottom of the workpiece by a gas flowing around the arc very similar to laser cutting.

focusing gas flow
protective gas
coolant flow
ignition
tungsten
electrode

work piece uncut   cut kerf

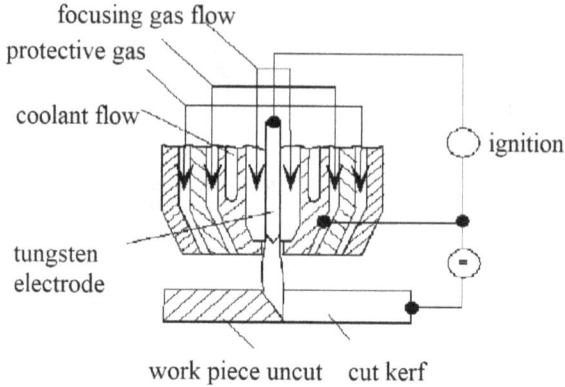

Fig. 11.4   Cutting head of a plasma cutter.

Source: D. Schuöcker, Spanlose Fertigung, Fig. 2.2.3, Oldenbourg 2006.

If a very narrow cutting gap is to be realized in order to reduce the total heat input, which is proportional to the volume molten per unit of time and thus also to the width of the cutting gap and the thickness of the workpiece, it is necessary to focus the arc on the surface of the workpiece. This can be done by the fact that the arc, which burns between a tungsten electrode and the workpiece, is surrounded by a cold gas flow at high speed (see Fig. 11.4.), which causes it to be additionally cooled on its surface, apart from the radiation, by the flowing gas, thus reducing the cooled surface. Such an arc is called a "plasma beam." If also a protective gas is used to avoid the oxidation of the workpiece, one arrives at a *plasma torch* in which a central tungsten electrode generates the arc. The nozzle surrounding this electrode generates the focusing gas jet and another nozzle coaxial to the first nozzle supplies the necessary protective gas. These nozzles are made of well-heat conducting copper and water-cooled to avoid heating too much.

If the plasma is ignited between the tungsten electrode and the workpiece, which must be electrically conductive, one speaks of a "direct plasma beam." If the workpiece is not electrically conductive, there is also the possibility to ignite the plasma beam between the inner tungsten electrode and the surrounding nozzle mouth and then deflect the plasma beam through the gas stream to the workpiece until the plasma comes into contact with the workpiece and therefore heats it.

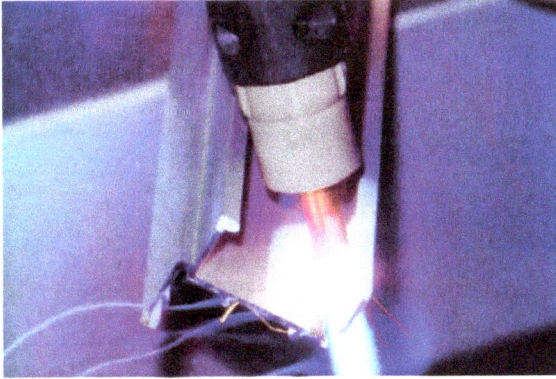

Fig. 11.5   Sheet metal cut by a plasma burner.

Source: D. Schuöcker, Spanlose Fertigung, Fig. 2.2.5 Oldenbourg 2006.

Such plasma burners can be operated with outputs up to a few 10 kW, where an argon-hydrogen mixture is used as focusing gas.

Besides cutting the plasma burner can also be used for welding and for removing paints on metal. The plasma jet can be shielded from the environment by a water sheath, thus shielding the dangerous UV radiation that each plasma produces as a result of the high temperatures. On the other hand, the noise of the gases emanating from nozzles and finally also removal products can be dissipated.

The maximum workpiece thickness of steel is about 100 mm and the maximum speed is 10 m/min for 1 mm thickness, decreasing at larger thickness. The cutting quality is determined by rounded cutting edges and a cutting kerf in the range of some millimeters that widens towards the plasma beam. Due to the strong heating of the work piece, heat-sensitive structures can be damaged.

Compared to laser cutting, the much larger workpiece thickness is an advantage, nevertheless the wide cutting kerf with rounded edges and the respective material loss are disadvantages. Overall, laser and plasma cutting complement each other very well.

11.2.5 *Water jet cutting*

Figure 11.6 shows the principal layout of a high-pressure water cutting system. It consists first of a hydraulic pump with pressure converter that

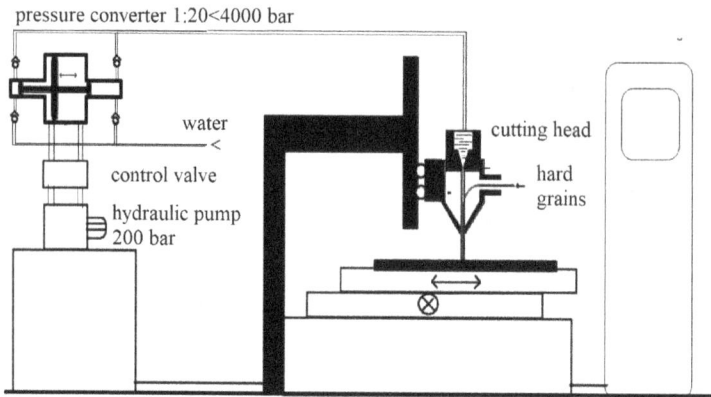

Fig. 11.6   High pressure water cutting machine.

Source: D. Schuöcker, High power lasers in production engineering, WSP and Imperial College Press, Fig. 5.5, Singapur/London 1999.

enhances the water pressure to 4,000 bar and feeds it to the cutting head.

A nozzle at the bottom generates a sharply focused water jet and directs it onto the workpiece. The cutting head is also supplied with a flow of abrasive material, as sand and mixes it with the water thus making the latter very abrasive. The work piece is carried by a conventional XY table and the whole system is controlled by a CNC. The mechanism of material removal (Fig. 11.7) is similar to that of laser cutting because erosion takes place at the momentary end of the cut kerf with the difference that material is not ejected as melt as in the case of laser cutting but in little solid chips that are split off due to the starvation pressure of the water jet acting on the momentary end of the cut.

The latter pressure leads to fatigue of a thin layer that is divided than in little chunks. These particles are then washed out due to friction with the abrasive water jet and leave the workpiece at its bottom.

Compared to laser cutting, at similar speed, the workpiece thickness that can be cut is considerably larger up to multiples of 10 mm, where steel is difficult due to its extreme strength, but aluminum and similar metals can be cut excellently. Unfortunately strong wear takes place at the nozzle owed to the abrasive means added to the water jet. Moreover the process is noisy. Nevertheless water jet cutting is a valuable addition to laser cutting since it extends its thickness range.

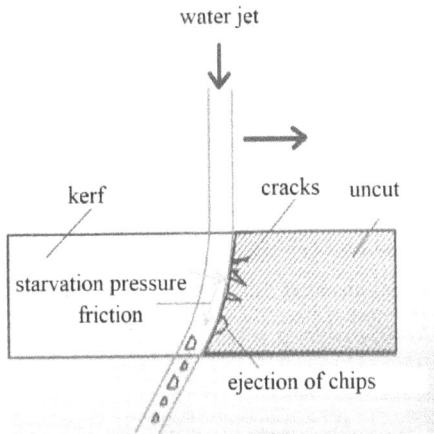

Fig. 11.7   Mechanism of water jet cutting.

Source: D. Schuöcker, High power lasers in production engineering, WSP and Imperial College Press, Fig. 5.8, Singapur/London 1999.

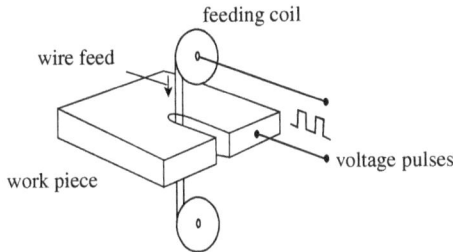

Fig. 11.8   Principal function of wire erosion.

Source: D. Schuöcker, Spanlose Fertigung, Fig. 2.1.6, Oldenbourg, München 2004.

Special advantages in cutting of plastics are that no heating of the work piece takes place and that no vapor or dust is emitted.

11.2.6 *Wire erosion*

If one does not want to produce 3D recesses with the help of spark erosion, but 2D cuts, one uses a thin wire (see Fig. 11.8), which serves as a tool electrode. Sparks ignite between the wire and the workpiece over its full thickness. Thus, the material is removed in the area of the projection of the wire onto the cross section of the workpiece and a cut with a width corresponding to the very small wire thickness is created. If the wire is now moved in two directions relative to the workpiece in a plane parallel to the workpiece surface, any contour can be cut out. Since the wire also suffers from erosion and thus tends to tear off due to the associated cross-sectional reduction, it is continuously unwound by a storage coil during the erosion process and wound back onto a second coil. With this wire erosion method, very thick workpieces can be cut up to several decimeters thickness with high quality, i.e., small cut kerf and low roughness, although the feed speed can only be very low. Typical wire thicknesses are 20–250 microns and cutting performances (v-d) are 10 mm2/min. Figures 11.8 and 11.9

Fig. 11.9   Wire erosion.

Source: D. Schuöcker, Spanlose Fertigung, Fig. 2.1.8, Oldenbourg, München 2004.

Fig. 11.10   Workpiece cut by wire erosion.

Source: D. Schuöcker, Spanlose Fertigung, Fig. 2.1.9 Oldenbourg, München 2004.

illustrates the removal process. Figure 11.10 shows a workpiece produced with it.

11.3 Ablative Processes

11.3.1 *Spark erosion*

In this process, a tool electrode is used, the shape of which corresponds to the recess to be made in the workpiece and wherein a certain wear by erosion as mentioned above must be taken into account in the design of this electrode.

This electrode is now moved by a linear feed device towards the work-piece surface and into the already manufactured recess (see Fig. 11.11), so that the bottom of the recess is lowered further and further into the work-piece. Typical removal rates for steel are 30 mm³/min ($I = 16$ A, $U = 150$ V, pulse length 100 s, $f > 6$ kHz).

Figure 11.12 shows an industrial erosion machine and Fig. 11.13 shows a workpiece produced with it.

In order to be able to produce more or less any recess geometry with-out specially manufactured tools, an electrode in the form of a thin, round steel can also be used (see Fig. 11.14), wherein this can then be moved both in the direction of the workpiece surface and parallel to it, thereby in both possible directions.

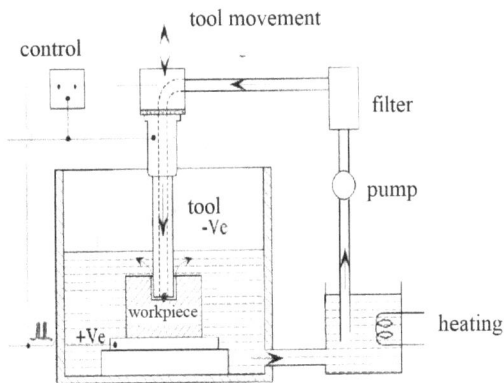

Fig. 11.11   Schematic view of a spark erosion machine.

Source: D. Schuöcker, Spanlose Fertigung, Fig. 2.1.2 Oldenbourg, München 2004.

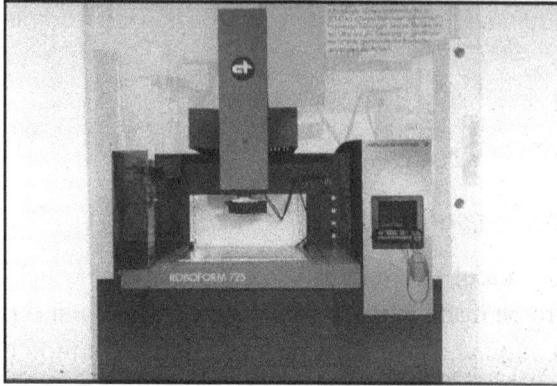

Fig. 11.12 Industrial erosion machine armed with sinking tool.

Source: D. Schuöcker, Spanlose Fertigung, Fig. 2.1.3 D. Schuöcker, Spanlose Fertigung, Oldenbourg, München 2004.

Fig. 11.13 Casting die produced by spark erosion.

Source: D. Schuöcker, Spanlose Fertigung, Fig. 2.1.5 Oldenbourg, München 2004.

The sparks then jump from the front side and from the cylindrical edge of the electrode to the workpiece. In order to avoid uneven wear of the circumference of the tool electrode, it is rotated around its own axis. The resulting total movement of the tool electrode is called "planetary motion."

Fig. 11.14   Erosion electrode with planetary movement.

Source: D. Schuöcker, Spanlose Fertigung, Fig. 2.1.4 Oldenbourg, München 2004.

11.3.2 *Milling*[1]

11.3.2.1 *Cutting, shear and chip formation*

The milling tool (Fig. 11.15), the cutting wedge, is moved with respect to the workpiece, thus acting on the material with a shear force, which leads to shearing if the force acting on the cutting wedge is sufficiently strong. Due to the wedge shape of the tool, the sheared material is now lifted by the rake face and flows through the forward movement of the tool along the rake surface away. The chip formed in this way is eventually hardened by the high temperature that develops through the shearing process under certain circumstances and breaks off at too large length, whereupon a new chip is formed. There is a free space between the flank of the cutting

---

[1]The following sections are based on Wikipedia, the free encyclopedia articles about milling, mills, grinding and polishing.

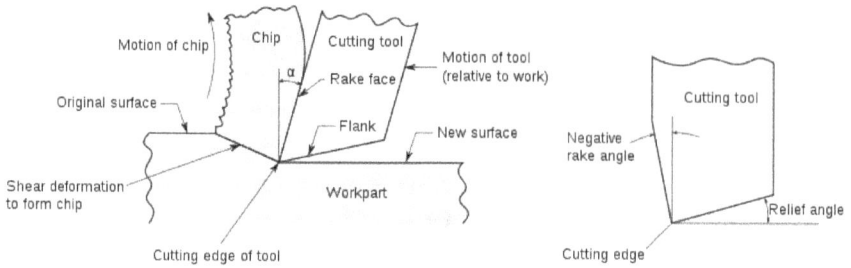

Fig. 11.15 Cutting wedge and material removal by chip formation.

(By Metal_Cut_diag.jpg: Sumanchderivative work: Swisstack — Metal_Cut_diag.jpg, CC BY-SA 3.0, https://commons.wikimedia.org/w/index.php?curid=11339746).

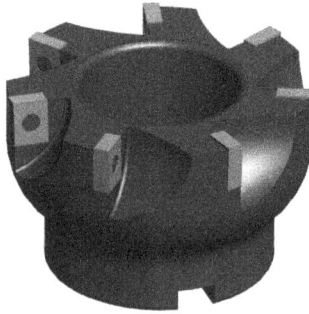

Fig. 11.16 Face mill

(By Rocketmagnet — Own work, CC BY-SA 3.0, https://commons.wikimedia.org/w/index. php?curid=2503447).

wedge and the workpiece, which prevents too much friction of the tool with the workpiece that would degrade the quality of the new surface.

11.3.2.2 *Milling tools*

There are two main types of cutting tools, the mills, the *face mill* with cutters where cutting edges are arranged at the circumference of its face and the *end mill* (Fig. 11.17) that carries cutters at the face and on the sides. The first one is used for milling plane surfaces or shallow pockets and the second is similar to a drilling bit, but cannot only be moved in

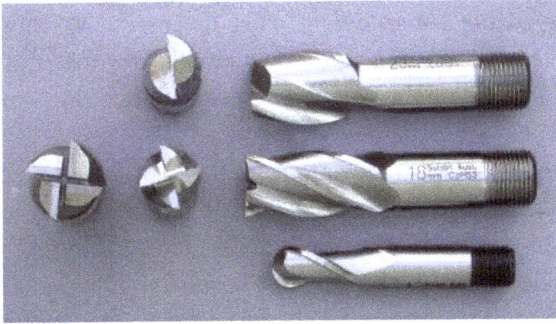

Fig. 11.17 End mills.

Fig. 11.18 Hobbing.

vertical direction but also perpendicular to its axis, thus cutting also walls that are in parallel to the axis of the mill and are thus well suited for the creation of deep pockets and similar geometries.

Special mills have a ball shaped head that can be used for milling corners with a certain radius. So called *hob mills* (Fig. 11.18) are used for the production of gear wheels, where the rotating hob is moved in parallel to the spindle rotating the workpiece and thus cutting out the gaps between the teeth. Many other types of mills are available for various applications.

Cutters are available for *roughing* what means quick material removal with poor surface quality and for *finishing* with slow material removal but yielding high surface quality. Usually more than one ride is necessary for roughing and also for finishing.

Usually the mills are made from HSS (High Speed Steel) and extra hardened or even coated with a hard cover to reduce wear. The most recent solutions are inserts made from carbide ceramics, for instance tungsten or titanium carbide (see Fig. 11.16). They have strong hardness and doubled lifetime but are quite brittle and are likely to break. They have usually more than one cutting edge and if the latter is worn out they can be turned to enable a new edge to come into action. If all edges are worn out, the insert is simply exhausted, where the conventional tools must be refurbished. To reduce friction and to cool the tools, a lubricant submerges the cutting edge.

Usually the rotational speed of the spindle moving the tool is 10–1,000 rot/min. The speed of the rotating tool with respect to the workpiece is 10 to 60 m/min for some steels, and between 100 m/min and 600 m/min for aluminum.

The depth of the cut, the thickness of the chip, is determined by the geometry of the tool.

11.3.2.3 *Material removal mechanism*

In milling of plane surfaces face mills as mentioned above are used, that carry a multitude of cutting tools as above and rotate around an axis perpendicular to the workpiece surface having wedges with horizontal edges extending in radial directions and removing ring shaped areas.

If in a certain moment one of the cutting wedges leaves the rim of the already generated trace due to its rotation, it starts to cut. After it has removed a half moon shaped layer it approaches the opposite rim of the trace and stops cutting that will begin again after a full revolution. Sometime after the beginning of cutting of the first wedge the next wedge begins to cut but due to the movement of the tool across the workpiece material is removed from a second trace that is neighboring the first one. So one after the other cutter starts to cut, what extends the area where material is removed. Therefore always one half of the wedges are active and the remainder relaxes, that means cools down after heating by friction.

The material utilization here is only about 50% due to the loss of material by chips. The energy required for machining is similar to the forming process of around 50–100 MJ/kg.

If roughening and subsequent finishing milling are used, high surface quality with a roughness below 1 μm can be reached.

11.3.2.4 *Movements of tools and workpiece*

Machining centers for milling have usually five axes of movement, whereby the workpiece can then be moved in three directions, which is often supplemented by a rotary table. A spindle for tool drive can be rotated around two axes so that the tool can be directed at any angle towards the workpiece. Also gantry solutions are used advantageously (see Figs. 11.19 and 11.20).

Usually also a tool changer (concourse) is provided. Very complex geometries can be realized. As the programming of such a machine is extremely complex, it is usually not done manually by the programmer but by a CAM program similar to 3D printing using slicer programs.

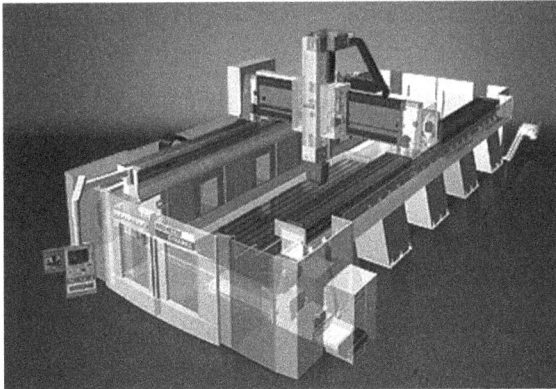

Fig. 11.19 Universal milling center with three linear movements.
(With kind permisionh of EMCO, Hallein/Austria, 2020).

Fig. 11.20　Universal tool drive with two rotational axes.
(EMCO Hallein/Austria, 2020) With kind permisionh of EMCO, Hallein/Austria, 2020).

11.3.2.5 *Suitability of metals for milling*

Steel as most important metal with less than two percent carbon is very well suited for machining, especially as *construction* steel. *Cast iron* with more than two percent carbon is usually milled after casting and performs quite well. There is only one exception if cementite is present. If the cast contains ferrite or pearlite where carbon parts separate iron, good milling results are obtained and even preferable lubrication by carbon takes place.

Hardened steels cause strong tool wear and frittering of the chips.

Stainless steel with martensitic structure asks for strong cutting forces due to its high hardness.

Aluminum that is usually machined to manufacture finished parts is soft and shows thus some smearing of edges. Tools are preferably diamond or tungsten carbide, titanium based inserts should not be used.

Copper forms a hard skin after casting that is difficult to mill, although the bulk material makes no problems.

Titanium is also fine, but if dust is formed it is dangerous since it enflames at 33°C.

11.3.2.6 *Equivalence of milling and 3D printing*

Practical data describing milling can be found in the internet and are shown as an example in the following table (see Table 11.1).

They are confronted with data of 3D processing with the powder bed process. Although the latter applies to material addition, it can be compared to milling, a material subtraction process, since both are used to generate 3D parts. The table shows clearly that the building rate of milling by removing the material is much higher than the building rate of 3D printing at comparable power consumption. The roughness obtained with milling is also much better than that obtained with 3D printing. Nevertheless material is wasted in the case of milling and not shown in the table, printing is much better for the environment since milling generates an enormous mass of chips that must be cared for. Also 3D printing avoids any tool wear, a critical point in milling.

11.3.3 *Grinding*

11.3.3.1 *Overview*

Grinding is carried out with a tool, the *grinding wheel* that consists of hard particles, embedded in the host material, a *resin*. Grinding wheels can have various shapes quite similar to milling, for instance plane circular geometry similar to a circular saw tool or finger like or cone shaped. These tools are used in grinding machines, where the most recent products are five axis machines similar to milling centers.

11.3.3.1.1 Abrasive grains

Aluminum oxide is used to grind ferrous metals. *Silicon carbide* is used for grinding softer, non-ferrous metals and high-density materials, such as cemented carbide or ceramics. *Boron nitride* or "CBN" and diamond, so called super abrasives, are also used. Grain size matters, since large grains remove material faster and smaller grains allow finishing.

11.3.3.1.2 Binders

The binders hosting abrasive grains are vitrified bonds, glass-like materials formed of fused clay or feldspar. Organic bonds as synthetic resins,

Table 11.1   Comparison of milling and 3D printing.

Steel			Milling	3D printing
Rotational speed	n		2654	
Feed	f	mm/min	637	
Cutting speed	vc	m/min	100	
Number of cutting edges	z		4	
Diameter of tool	d	mm	12	
Feed constant	fz	mm	0	
Depth of milling	tm	mm	30	
Force per unit area	F	N/mm^2	2800	
Chip width	h	mm	1	
Machined volume p.unit time	dV/dt	cm^3/min	229.30	1.5
Energy per unit time	P	kW	10.70	0.4
Energ per unit volume		J/cm^3	0.05	130
Roughness		μm	>0.1	<50
Material use		%	>50	100

Source: Authors.

rubber, or shellack are also used. Wheels that hold the abrasive grains strongly wear less, but show dulled grains resulting in bad grinding quality. In contrary softer wheels wear quickly.

11.3.3.1.3 Shapes

Various shapes include straight geometries similar to a saw blade (see Fig. 11.21) and also cup or saucer like geometries as well as conical designs and cannot only be used to grind flat areas but also support the creation of shapes. Cylindrical wheels provide a large, wide surface with no center mounting support (hollow). They can be very large, up to 12″ in width.

Cylinder or wheel ring is used for producing flat surfaces, the grinding being done with the end face of the wheel.

Diamond wheels are grinding wheels with industrial diamonds bonded to the periphery. They are used for grinding extremely hard materials.

Fig. 11.21   Disk shaped grinding wheel.

(CC BY-SA 2.0, https://commons.wikimedia.org/w/index.php?curid=90185).

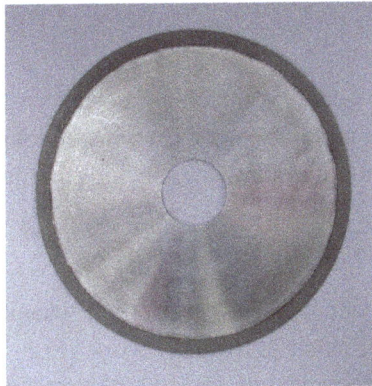

Fig. 11.22   Diamond wheel.

(CC BY-SA 2.0, https://commons.wikimedia.org/w/index.php?curid=90186).

Cut off wheels are self-sharpening wheels that are thin in width and often have radial fibers reinforcing them.

Grinding produces sparks and little fragments of metal, called *swarf* (Fig. 11.23), a typical characteristic.

Lubrication by use of fluids in a grinding process is often necessary to cool and lubricate the wheel and workpiece as well as to remove the debris produced in the grinding process. The most common grinding fluids are water-soluble chemical fluids, water-soluble oils, synthetic oils, and petroleum-based oils.

Fig. 11.23   Hand held grinding machine.

(By Lacholazarovphotos — Own work, CC BY 4.0, https://commons.wikimedia.org/w/index.php?curid=64390517).

Coolants reduce grinding machine power requirements, maintain work quality, stabilize part dimensions and insure longer wheel life. In use are emulsions, synthetic lubricants or special grinding oils.

11.3.3.2 *Mechanism of material removal*

The abrasive grains contained in the grinding wheel act as cutting tools, removing tiny chips of material from the work. (Fig. 11.24) As these abrasive grains wear and become dull, the added resistance leads to fracture of the grains or weakening of their bond. The dull pieces break away, revealing sharp new grains that continue cutting.

Material removal rate is 0.1 cm³/s, roughness 0.1 μm. The tolerances that are normally achieved with grinding are 5 μm for grinding of flat material.

11.3.3.3 *Types of grinding*

The following major industrial grinding processes are in use:

1. *Cylindrical grinding*
2. *Internal grinding*
3. *Surface grinding*

Fig. 11.24 Hard grains in the grinding wheel remove irregular small chips.
(By Jahobr — Own work, Public Domain, https://commons.wikimedia.org/w/index.php?curid=3092801).

In cylindrical grinding, the workpiece rotates around a fixed axis and the traces machined are concentric to the axis of rotation. Cylindrical grinding produces an external surface that may be either straight, tapered, or contoured.

Internal diameter (I.D.) grinders finish the inside of a previously drilled or bored hole, using small grinding wheels at high rotational speed.

Surface grinding uses a rotating abrasive wheel, creating a flat surface.

11.3.3.4 *Materials that can be grinded*

Advantageous use of grinding is obtained with cast iron and various types of steel. The latter materials can be held magnetically. Aluminum, stainless steel, brass, and plastics are also grinded but tend to clog the cutting wheel more than steel and cast iron. Grinding can also be used for cutting very brittle material, that cannot be cut with other means than lasers.

11.3.3.4.1 Grinding machine

Figure 11.25 shows a grinding machine as it can typically be found in metal workshops.

11.3.4 *Polishing*

Allows to improve the surface quality, e.g., after milling and grinding by the effect of a paste with submerged hard grains gliding under pressure over the

Fig. 11.25   Simple two axes surface grinding machine.
(CC BY-SA 2.5, https://commons.wikimedia.org/w/index.php?curid=304801).

surface to be improved. During this process very little material is removed in a microscopic thin layer. The paste is carried by a rotating disk that is pressed against the workpiece and moves across the latter. Beginning with a paste with coarse grains the grain size is subsequently reduced finally reaching very tiny grains, thus reducing the roughness from step to step, at the end resulting in a shiny metal surface, that shows practically no diffuse reflection, but similar to mirror geometrical reflection, where an incident beam of light causes a similar beam leaving the surface.

The paste itself consists of various oils and fats and hosts hard grains, e.g., from aluminum oxide.

Final roughness can become lass than 0.1 µm.

Grinding and polishing are not really competitors for laser processing but rather complementary, since, e.g., in 3D printing of metals with lasers, that result usually in a rather rough surface, that must be improved. For the latter task both grinding and polishing are perfectly suited.

Moreover polishing can also be performed with laser melting, thus avoiding manipulation of a workpiece generated by 3D laser metal printing and also change of tools as proposed very long time ago by the senior author and developed independently by ILT in Aachen/Germany.

Chapter 12

Competitive Manufacturing Processes with Material Addition

12.1 Welding of Metals

12.1.1 *Overview*

The most important non-laser manufacturing process with material addition is welding, that will thus be treated in the following.

Welding options can be distinguished by the kind of energy they use for melting the workpieces to be joined: as mechanical, electrical or chemical energy.

Since only electrical processes yield the most desired narrow weld seams comparable to laser welding, the actual chapter is restricted to arc and e-beam welding. The latter processes can be distinguished according to their mechanical setup, the gaseous environment and the kind of electrical power: as DC, pulsed or others.

Mechanical designs use welding wires fed to the workpieces to be joined continuously as electrodes, either one or even two. In some processes wires can also be drawn back besides being moved towards the workpiece. Also, fixed electrodes are used.

Current supply is determined by the current source that is subject to much progress with various temporal distributions as for instance DC or pulsed and others.

Chemical properties include inert or active-reactive gases.

12.1.2 *Gas Metal Arc Welding (GMAW)*

A GMAW welding burner uses a welding wire fed through a water cooled nozzle. A protective inert gas or an active-reactive gas flows around that nozzle and prevents hot metal from oxidation or supports the process by exothermal reaction and adding desired materials to the weld seam. The wire and the workpiece are connected to the current source and act as electrodes for the arc. Ignition of the latter takes place if the wire is moved towards the workpiece and touches it, thus causing a short circuit with very high currents that, due to strong heating, lead to melting and evaporation of the wire that interrupts the contact between wire and workpiece. Thus, a gap is formed that is filled with metal vapor that becomes ionized due to the high temperature of evaporation and allows thus the continuation of current flow as an arc. If now the wire is moved away from the workpiece, the arc elongates what causes the voltage across the arc to rise, quite similar to the voltage across a resistor that rises with rising length of the latter. Eventually, the voltage across the arc reaches the voltage driving the current source, thus limiting the length of the arc. The latter with its very high temperature (up to 5000°C) heats on the one hand the workpiece up to the melting point and melts thus the two work pieces to be joined. On the other hand, the arc heats the tip of the wire up to melting, leading to the formation of a molten droplet, that grows continuously until it is heavy enough to take off from the wire. Subsequently the drop moves towards the workpiece and contributes to the filling up of the gap between the two members of the joint (see Fig. 12.1).

Accordingly the two main parts are the *nozzle* (equipped with water cooling, *wire feed* and *gas flow*) and the *current source*, the latter providing many forms of currents: DC or pulsed or of more complicated nature. It also allows to perform the developments such as TIME or CMT processes (see below).

The wires used in GMAW welding are similar to the composition of the metal being welded and contain deoxidizing metals such as silicon, manganese, titanium and aluminum in small percentages to form slag that protects the hot weld seam and helps to prevent oxygen porosity. Some contain denitriding metals such as titanium and zirconium to avoid nitrogen porosity. Depending on the process variation and base material being

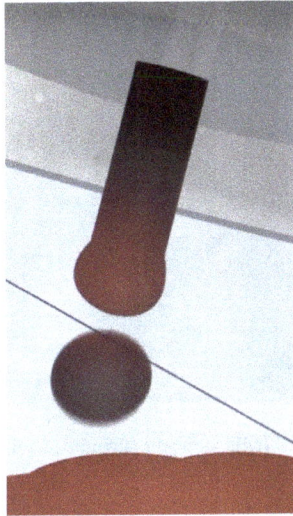

Fig. 12.1   Droplet formation, detachment and motion in arc welding.
With kind permission by Fronius, Wels/Austria).

welded the diameters of the electrodes used in GMAW typically range from 0.7 to 4 mm with a feed rate of up to 50 m/min.

Shielding gas mixtures of argon, CO_2 and oxygen are used for steel and argon or helium for nonferrous metals with a consumption of 20 l/min. Welding speeds range up to several 10 m/min.

T.I.M.E (Transfer Ionized Molten Energy): Welding uses a longer free part of the wire after leaving the nozzle and a gas mixture (of inert and active gases), mainly two wires in a common nozzle and gas flow (TIME twin as in Fig.12.2).The latter are isolated from each other and are independently supplied with current, for instant asynchronously, that means that drop formation and motion takes place always at only one wire and then at the second one. With the latter arrangement the welding speed can be doubled, a major advantage.

CMT (Cold Metal Transfer): In this process, the welding wire under voltage is moved in the direction of the base material until a short circuit forms. After adjusting the current flow, the power supply is interrupted (computer-controlled) and the welding wire is moved back in the

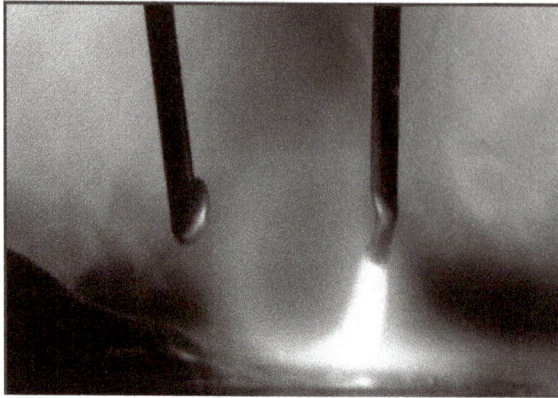

Fig. 12.2   TIME welding (Fronius, Wels/Austria).

With kind permission by Fronius, Wels/Austria).

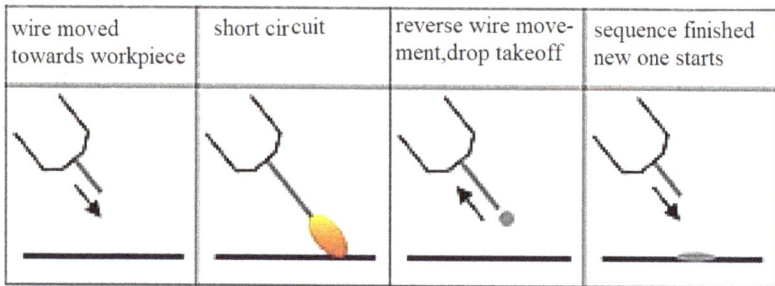

wire moved towards workpiece	short circuit	reverse wire move-ment,drop takeoff	sequence finished new one starts

Fig.12.3   Sequence of steps in CMT welding (Fronius, Wels/Austria).

With kind permission by Fronius, Wels/Austria).

opposite direction. Due to the wire movement, the molten droplets form-ing during the short circuit are particularly easily detached from the wire. Due to the controlled power supply and the supporting effect of the wire movement during the material transition, only a very low heat input to the base material is carried out. Advantages of this are a small heat-affected zone.

Cold arc welding: Cold arc welding uses a short arc in which material transfer is obtained by a relatively low current with less heat input sup-ported by a pulse that facilitates melting.

12.1.3 *Tungsten Inert Gas welding (TIG)*

This process uses a fixed tungsten electrode that does not melt and is thus not consumed. It is similar to GMAW: surrounded by a water-cooled nozzle that guides a flow of inert protective gas, either helium or argon to then workpiece. A filler wire is supplied laterally from outside. The quality of the weld seam is very good; that means crack free and without pores. Nevertheless, the welding speed is lower than in GMAW welding. It is used mainly for welding stainless steel, aluminum, copper and similar materials.

12.1.4 *Plasma welding*

This process is of special interest for the comparison to laser welding since it approaches the latter due to a relative small footpoint on the work piece.

The latter is caused by an extra flow of gas, the plasma or focusing gas around the arc, that cools it at its circumference. This cooling disturbs the energy balance of the arc that loses energy at the two footpoints and by radiation from the column, decreasing with decreasing radius of the latter. If now additional cooling of the column takes place due to the cooling focusing gas, at constant heat gain by electrical energy the radiation loss must he reduced by a decreasing column radius (Fig. 12.4).

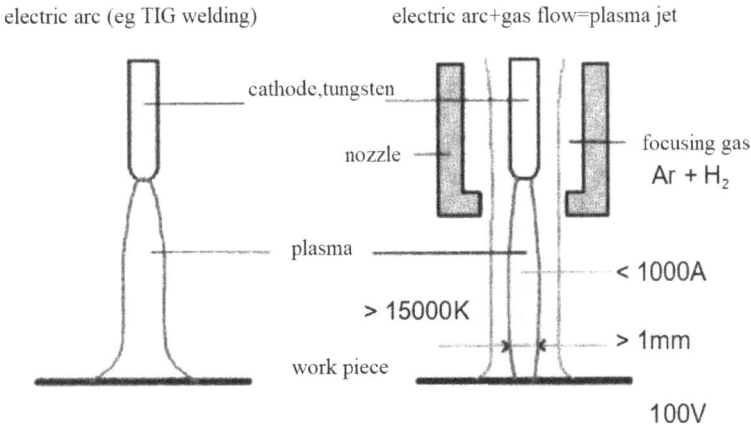

Fig. 12.4 Focusing an arc by a cooling gas flow 2.2.1.

Source: D. Schuöcker, Spanlose Fertigung, Fig. 2.2.1 Oldenbourg, München 2004.

To use the mechanism of plasma gas cooling and thus focusing the arc on the work piece, the water-cooled welding nozzle has a central opening for the non-consumable tungsten electrode and the plasma gas flowing around the latter and a concentric opening for the protective inert gas (see Fig.11.4).

12.1.5 *E-beam welding*

E-beam welding is based on the presence of numerous free electrons in a vacuum environment that are deliberated from a metal, the cathode, due to high temperature. That means a high kinetic energy, allowing to escape from the attraction by the positively charged atoms in the metal. These electrons are attracted by a positively charged electrode, the anode, and enter finally the latter thus generating a current flow driven by a voltage applied to the two electrodes (Fig. 12.5). The voltage mentioned before can be made very high in the order of magnitude of many kilovolt and so the electrons get high speeds in axial direction, thus prevailing any lateral movement. Moreover, electric fields perpendicular to the current flow created by auxiliary electrodes act on the electron flow and compensate their mutual repulsion. All these measures grant a very low divergence of the electron beam (Fig. 12.6). The electrons hit the anode with high kinetic energy in a very small spot, thus heating it up quickly to the melting and even evaporation point, allowing use for welding with the advantage of

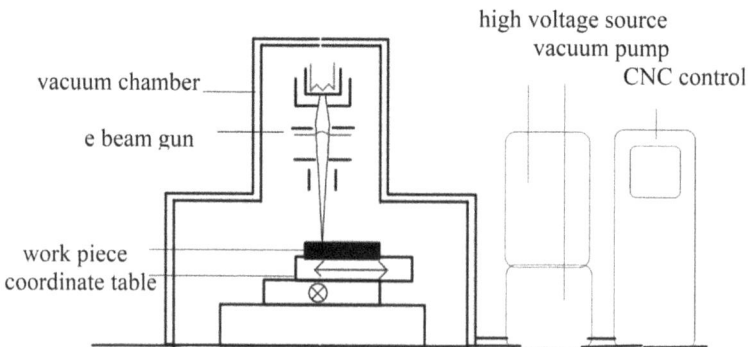

Fig.12.5   E beam welding system.

Source: Authors.

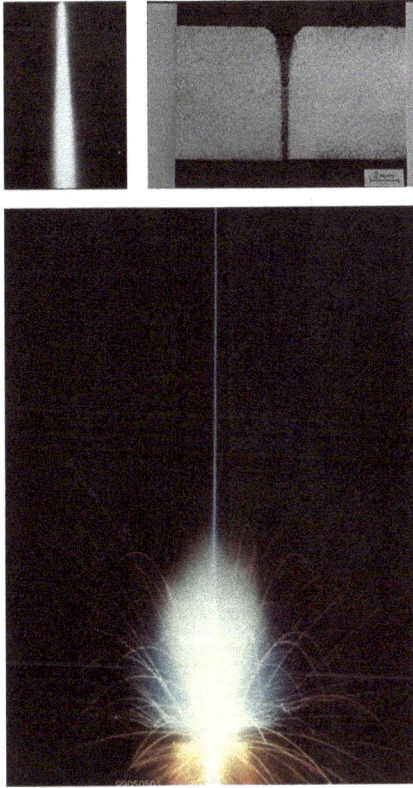

Fig. 12.6 E beam Von Doc Klauser — Eigenes Werk, CC BY-SA 4.0, https://commons.
wikimedia.org/w/index.php?curid=45215899

high penetration depth up to multi 10 cm, and also a narrow seam
(Fig. 12.7) that ensures low heat input to the work piece; all that based on
the very low divergence. A disadvantage is the necessity of a vacuum
environment (Fig.12.5). Nevertheless, e-beam welding has already been
performed without vacuum, although with increased divergence due to
collisions of the electrons with air atoms.

12.2 WAAM Wire Arc Additive Manufacturing

This process is based on a non-laser 3-D printing system, in which a wire
is molten down on a workpiece under construction layer-by-layer by an

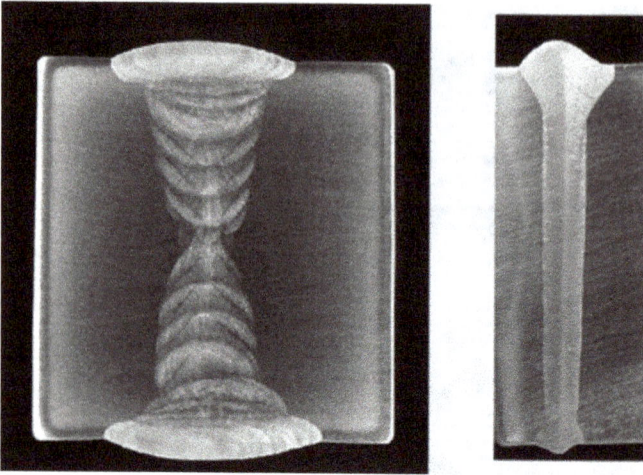

Fig. 12.7   Seam cross section of an e beam weld copareed to arc welding (30 m)

Von Doc Klauser — Eigenes Werk, CC BY-SA 4.0, https://commons.wikimedia.org/w/index. php?curid=42687258

Fig. 12.8   WAAM with robot guided welding head.

Source: (IGM, Wr. Neudorf/Austria, 2020).

Fig. 12.9   Creation of a steel layer by welding down a wire with a CMT arc.
Source: IGM, Wr. Neudorf/Austria, 2020.

electric arc achieving high build rates and good quality (WAAM; Wire Arc
Additive Welding) (Fig. 12.8 and 12.9). Arc welding is performed by a
CMT welding process (see above) because it results in minimum heating
of the already manufactured layers. This new process is favorable in view
of investment and operating costs compared to similar laser processes,
which can be replaced by the new solution. Further development of
WAAM could also reach higher build rates and yield a breakthrough in 3D
printing of metals.

Chapter 13

Competing Technologies without Mass Change

13.1 Heat Treatment Overview

Usually metals are finally obtained from a molten phase by cooling down. If cooling takes place slowly, crystallites have time to grow and big grains result that make the metal soft and weak (see Chapter 9). If cooling is fast only small grains can be formed and give the metal more hardness. Nevertheless the microstructure can be changed by heating the metal up to the annealing temperature where the grains are dissolved and vanish. New grains can now be formed by quick or slower cooling thus yielding the desired hardness. Due to these considerations heating to a certain extent and cooling with different speeds can alter the material properties considerably.

The most important application is hardening of metals, where either the bulk or only the surface can be treated. Bulk hardening (Fig. 13.1), can be performed either by *transformation hardening* with changes of microstructure or by *precipitations* that disturb the microstructure or finally by *work hardening* with deformation of the grains due to forming. Heating can be performed in a furnace or a plasma and finally by inductive heating. *Case hardening* that keeps the bulk of the workpiece tough and ductile can be performed by *nitriding* that distorts the lattice at the surface, creating stress and *carburizing*, that enhances the carbon content, thus

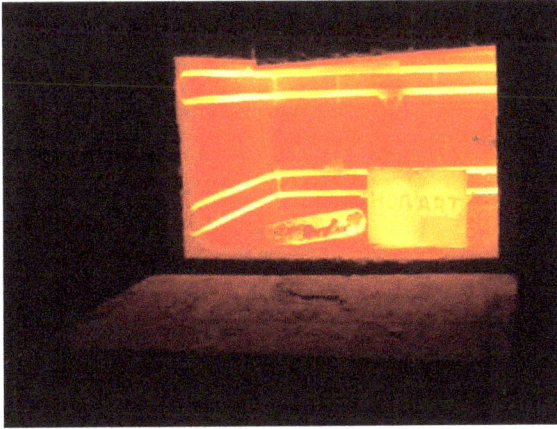

Fig. 13.1   Furnace for workpiece hardening.

allowing transformation, can be carried out by heating in an appropriate atmosphere.

13.1.1 *Transformation hardening*

Has already been treated in Chapter 10 and applies to carbon steel with less than 0.8% carbon that changes its lattice structure at the transformation point (723°C). If the material is then rapidly cooled down, the lattice changes again to that of ultra-hard martensite:

Initially at room temperature *ferritic* steel has a body centered microstructure, where one iron atom sits in the centre of a cube with iron atoms at the corners. At a temperature of 726°C the crystallization mode changes to face centered *austenite* with iron atoms in all corners and all face centers. In the case of carbon steel the carbon atoms are perfectly dissolved in iron and occupy interstitial sites but after quick cooling with the formation of a new body centered lattice, they have now time to carry out the necessary rearrangements and cause thus lattice defects that are associated with internal stress, thus forming very hard martensite.

13.1.2 *Work hardening*

Is obtained after deformations of a metallic workpiece since in this process the grains also undergo deformations as elongation in one direction and shrinking in the other direction thus reducing the susceptibility for a movement of lattice defects that cannot jump over the grain borders and therefore obstructing deformations and enhancing hardness.

13.1.3 *Precipitation hardening*

Is obtained if small volumes with composition and microstructure different from the main material are embedded in the latter. These zones obstruct the movement of lattice defects. Primarily this mechanism works at elevated temperatures, but can also be carried out at room temperature, but then a long time is necessary for the formation of precipitations.

Although the first two hardening mechanisms are mainly used for steel (and titanium), the last one applies especially to aluminum.

13.1.4 *Case hardening*

Concerning surface hardening, nitriding builds-in nitrogen atoms in the surface of the workpiece and exerts stress on the regular lattice, thus leading to enhanced hardness. It can first be performed with a hot workpiece heated in a furnace and exposed to an atmosphere of NH_3. The latter gas is then dissociated due to the high temperature and atomic nitrogen is absorbed by the workpiece surface. Also the cold workpiece can be submerged in a bath of salt containing nitrogen.

Finally the initially cold workpiece can be used as cathode (negative electrode) of a nitrogen plasma, either glow or arc type, that contains positively charged nitrogen ions attracted by the negatively charged workpiece and thus entering the latter and enriching it with nitrogen with a hardening effect as above.

Also carburizing can lead to increased hardness, since it enhances the carbon content in the workpiece surface and thus making transformation to martensite possible. The latter process is performed by heating the workpiece and submerging it in a carbon dioxide atmosphere, in which the latter gas dissociates due to the high temperature of the part to be treated. Therefore atomic carbon is deliberated and settles at the surface of the workpiece thus enhancing the carbon content of the latter.

13.2 Annealing

Annealing means heating of a workpiece to a temperature above the recrystallization point, holding it a certain time at this temperature (some minutes or more) and then slowly cooling down to room temperature mainly in order to remove hardness and stresses.

13.2.1 *Stress relief*

Intrinsic stress in a workpiece is a stress that is not caused by external forces. It appears especially in welding, in which the liquid state bridges

Fig. 13.2   Casted parts after heat treatments.

Von Goodwin Steel Castings — Flickr: Castings fresh from the heat treatment furnace, CC BY-SA 2.0, https://commons.wikimedia.org/w/index.php?curid=16070140.

the two parts. When cooling starts, the melt shrinks during resolidification, while the majority of the two workpieces are relatively cold and therefore do not participate in the shrinkage, which in the cooled seam induces stress, which may also lead to cracks (Fig. 13.2).

Stress relief annealing of steel is made at a temperature of around 600°C by a long time (6 h) arrest of the temperature rise and a very slow cooling (14 h), since during the arrest the high temperature reduces the yield strength to become smaller than the internal stress thus causing yielding and elongation of the seam that matches the cold rims thus eliminating the internal stress to a high degree, but not totally.

13.2.2 *Normalizing and tempering*

Normalizing of steel is obtained by heating it up to a temperature a bit higher than the critical temperature for austenitization and after some time storing it in ambient air. Normalized steel consists of pearlite, martensite and sometimes bainite grains (see Chapter 9) mixed together within the microstructure. The steel is then much stronger than full-annealed steel and much tougher than **tempered steel**, that means steel kept at a temperature of around 200°C for some time. However, added toughness is sometimes needed at a reduction in strength. Subsequent tempering increases the toughness to a more desirable point. Cast-steel is often normalized rather than annealed to decrease the amount of distortion that can occur. Tempering can further decrease the hardness, increasing the ductility to a point more like annealed steel. Tempering is often used for carbon steels, producing much the same results.

13.2.3 *Recrystallization annealing*

Removes work hardening by dissolving the deformed grains and growing new undisturbed ones with low carbon steels. It is carried out at temperatures between 400°C and 700°C during a holding time of more than 10 hours.

13.3 Heat Treatment Furnaces

Industrial furnaces allow heating up to 1,600°C and also chilling in different atmospheres as vacuum, air, hydrogen and many others for the purpose of bulk hardening and case (surface) hardening or annealing, for controlled reduction of hardness, stress relief and others. They can work continuously with a band transporting the workpieces to be treated along a distance of several 100 m through the hot atmosphere or they can treat batches, a multitude of parts, at a time with a bell like arrangement, where a cap is lifted, the parts to be treated are loaded and then the bell like cover is lowered and closes the processing chamber. Heating is preferably carried out electrically.

Chapter 14

Machine Safety

Manufacturing a part means destruction of shape or structure of the base material followed by building new shape or structure, all processes that need excessive means, as strong forces, high voltages or elevated temperatures. All these incorporate **hazards for the operating personal**, where mainly injuries of eyes by split off particles in machining or light, especially in the UV range in arc welding, cutting of fingers or more by sharp and powerful tools and breaking of bones by quickly moving and heavy machine parts, especially with robots. Also electrical strokes by spark erosion machines that can lead to instant death and burning of skin and hair by electrical arcs or in furnaces hurting very much, must be expected under unsafe conditions.

Also robots due to their high mass and fast movement and the possibility of erratous programming can be dangerous, even fatal, what has been demonstrated years ago at a leading robot producer in Germany.

General measures to prevent all these risks are fast emergency interruption of all machinery and enclosure of manufacturing machinery either by solid fancies or by installations that register automatically the intrusion of persons and subsequent emergency switch off. An example are light beams extending around a dangerous machine and registering interruption.

These hazards and respective protection will be discussed in the following, again distinguished according to the nature of the production process, either material removing or adding and finally without mass change.

Material removal by non-laser cutting of sheet metal can first be performed with shearing tools, rotating saw blades or nibbling stamps. These machines are all dangerous since split off particles can injure eyes and fingers, even hands can be cut off. To avoid accidents, safety goggles must be worn and measures to prevent hands from gripping into machines during operation, as mechanical barriers or light beams that stop operation at interruption. Also machines that afford both hands can be helpful. Safety housings for processes using cutting arcs and sparks can be used to prevent hazards of UV radiation that can be harmful for the human eye and high tensions that can even be fatal, not to forget very high temperatures that can lead to strong pain if burning skin. Again safety goggles especially for welding are ultimately necessary, also protective gloves and fire safe clothing must be used. To exhaust fumes that can be emitted usually during melting and evaporation of metals, respective air removal and cleaning systems must be used.

Third, high pressure water jets can be used as cutting tool and can do harm to the personal if a part of the enclosure of the high pressure system loosens, since then water with 3,000 bar can form a jet that can injure humans severely. The latter risk can be reduced if the whole water cutter is surrounded by a fence from stable steel grid.

As far as it concerns non-laser ablation, many of the above threats and countermeasures apply equally.

The hazards of **processes with material addition**, practically welding, are much the same then for material removal with the exception that the boiling melt present during welding splashes liquid metal around thus making burning likely and can only be made less critical if the welders wear full body protection clothing.

Hazards of **processes without mass change** are similar to that in the first two categories and with specific importance of the risks associated to high temperatures and strong forces and respective countermeasures.

Chapter 15

Comparison of Conventional Technologies with Laser Manufacturing

Most laser processes can only be applied to sheet metals with a limited thickness, since laser radiation cannot penetrate much into the bulk of a metal due to absorption near the surface and due to the limited Rayleigh length, that is the distance from the focus where the beam remains collimated, that leads to spreading up of the beam. Examples are cutting, welding and hardening as well as forming with laser assistance that are all restricted to thin (<20 mm) sheets of metal or to the vicinity of the workpiece surface, as in hardening.

Therefore a most important question if comparing lasers with other technologies concerns the **maximum thickness** that can be treated with a distinct alloy. In this respect most conventional processes as of mechanical or electrical nature are preferable, since thick bulky materials with much higher thicknesses (up to several 100 mm) can be processed.

A further question concerns the **speed of processing**, a parameter that is in some processes comparable to non-laser processes, as cutting by nibbling and spark erosion or welding by electron beams or plasmas.

Furthermore the ability of lasers to process along nearly **arbitrary contours** is a very important advantage in the case of cutting. Of course the question after the **quality** of the processed workpiece in terms of fidelity to shape and dimensions, the roughness of the processed surfaces and residual stress and deformations, also softening or hardening must be considered. Also **tool wear** must be regarded. Finally material and energy **consumption** must be taken in account, where laser processes often are superior to conventional manufacturing, as high efficiency of material use in 3D processing and in some cases in welding, where no filler wire is needed. Also distortion of the **environment** must be regarded, as e.g., in cutting noise emitted in nibbling compared to silent laser cutting. Also in surface hardening lasers are superior since selectively distinct regions of the workpiece surface can be hardened thus avoiding hardening of the whole surface. Finally **safety** risks must be considered.

Due to these considerations, to finally judge on the primacy of lasers or competing technologies, the various processes must be compared in detail as in the following.

Laser cutting can be compared to those processes that allow to cut arbitrary contours, as nibbling, water jet cutting, plasma and spark erosion cutting. It is clearly superior to nibbling, because the latter is noisy as mentioned above and provides a saw too like cut surface that needs postprocessing, two disadvantages that cannot be compensated by lower investment. Water jet cutting suffers from strong wear of the nozzle, especially by friction with the necessary abrasive powder added to the water jet and therefore high operation costs, but is rather a completion than a competitor since it can cut much larger thickness. The same is true for plasma cutting and also for wire erosion, the last one able to cut in the multi 100 mm range although considerably more expensive similar to the high pressure system of water jet cutting. Nevertheless wire erosion is quite slow compared to laser processing. Thus it is the opinion of the authors that lasers combined with plasma cutting is a good solution in terms of speed and thickness, yields good quality at reasonable investment where the high costs of the laser sources are compensated partly by the much cheaper current source (a welding source costs €1,000–€2,000 and a laser for cutting with say 5 kW beam power 100 times more). Nevertheless for the most important thin sheets of steel, stainless steel and aluminum with 1–3 mm thickness, laser cutting is the ultimate solution.

In **grooving and engraving** the problem of thickness is not relevant and also the investment is much lower since only little laser power is necessary and thus lasers seem the best choice for these applications.

Laser ablation suffers from the problem that the workpieces must be processed line by line and the latter are narrow, much narrower than the traces of milling tools, that allow considerably higher speeds in the order of magnitude of 100 m/min. Moreover milling tools can create also vertical walls and even work inside hollow cavities or parts, options that are not available with lasers, that do not have the large diversity of processing due to the manifold of milling tools that reach from simple mills for surface removal comparable to laser ablation to sophisticated tools for manufacture of tooth wheels, as e.g., used by the company Zörkler of Austria where one of the authors heads the production. The latter process of milling is even more attractive due to a much lower investment and even much better surface quality that can reach a roughness of tenth of a micrometer.

Laser welding is strongly related to its competitors, arc welding and e-beam welding, since all three processes use heating of a small spot on the workpiece surface to perform the necessary melting of the two members of the joint. Nevertheless in the case of arc use, the area of the hot spot is by two orders of magnitude larger and also radiation going out from the arc plasma heats the workpiece in a much larger area around the spot, thus melting not only the weld zone but also a big part of the workpiece, what leads to stress and deformation. To avoid the latter both workpieces to be joined must be fixed together carefully. The workpiece thickness that can be reached in these processes is determined by the formation of a keyhole that allows heat to penetrate into the depth of the material. Only e-beam welding allows to reach much larger depths in the order of magnitude of several 100 mm due to the very low divergence of the beam. Nevertheless arc welding can also process thick sections up to several 10 mm with the help of a V shaped gap between the two parts that is then filled layer by layer with molten wire, a process that reduces the welding speed considerably.

Usually the quality of the laser made weld seams in terms of smooth surface and narrow cross section is excellent and the threat of pores and cracks can be reduced by the use of additional welding wire, especially to avoid pores in aluminum seams. Unfortunately the investment for the laser equipment is much higher than that for the welding current source, quite similar to e-beam welding in a vacuum chamber with an e-beam gun as well as a high voltage source. The latter process is not really a competitor for the laser but completes it for thick workpieces. All three processes emit hard UV radiation, in the case of the e-beams even X-rays and must thus carefully be shielded. Concluding arc welding that rules nowadays the joining business is and will be superior to e-beam welding and thick section laser welding will only be used for special materials, as conventionally only difficult to weld or for structures that may not become exposed to stress and deformations.

For laser welding of thin sections with a thickness in the order of 1 mm more simple heat conduction welding can be used. That process can be performed with relatively low beam power of 1 kW and a relative large focus that bridges the gap between the two parts. Therefore cheaper diode lasers can be used and thus it is superior to arc welding especially since the latter enhances the risk of deformations due to the low thickness of the material.

Special process options as hybrid welding as a combination of laser and arc as developed by Fronius of Austria, world leading in electro welding, allows to enhance the performance of pure arc welding by a factor of two.

Cladding with arc welding is a powerful possibility not only for the creation of surface layers but also for 3D printing of metals especially with robots that allow to generate even very large part. IGM of Austria, a robot producer has adopted and investigated successfully the latter process that could become a serious competitor for laser printing of metals, although lasers allow to build up more subtle structures with less surface roughness due to the much smaller heated area. Therefore arc driven printing asks definitely for post processing although the investment is much lower as stated several times. By the way, in 3D printing of plastics, experiencing enormous interest and growth of sales, the nessecary melting of the base material nowadays is solely done by electrical heating. With the use of lasers for heating the amount of mass molten per unit time could considerably be enhanced, thus improving the processing speed, that is at the moment much too low for industrial series production.

Finally every day welding at least for thick sections, will be done with lasers now and possibly also in the future only in niches.

Laser hardening is case hardening which affects only the surface of the workpiece and its vicinity. It enhances corrosion and wear resistance but leaves the bulk of the workpiece tough to withstand heavy loads. Case hardening is carried out in furnaces that use for heating up, keeping the temperature and finally cooling down much time and also much energy. Laser hardening can now been done without any furnace selectively only in those regions of the workpiece surface where corrosion or wear resistance are necessary thus saving time and energy. Thanks to a development at the Gmunden Laser Center, the quality of the hardened regions has been considerably improved by a temperature control, that avoids overheating at protrusions or holes that lead to melting due to obstructed heat conduction. The advantages of laser hardening apply to single pieces or small batches where for large batches furnaces that allow treatment of many parts at a time are preferable.

Laser assisted forming has the big advantage, that brittle materials, as e.g., Magnesium or Titanium can be formed without getting cracks or even rupture and that the amount of deformation can be largely increased for all metals due to softening work piece regions hardened by prior deformations (work hardening), thus allowing further deformation.

As a **result of the above considerations**, for metal manufacturing lasers in the multi kW range are essential, since they can be used advantageously for cutting, thin section welding, selective hardening and forming of brittle and hardened materials. Lasers even of very low power are perfectly suited for engraving all kinds of material. It can also be expected that they will more and more be used for 3D processing, in the near future also in plastics processing. Welding of thick section metals will remain a niche application, but case hardening with temperature control will be used to a rising amount for small series.

Index

www.ingramcontent.com/pod-product-compliance
Lightning Source LLC
Chambersburg PA
CBHW050600190326
41458CB00007B/2116